AF361922

High-Performance Green Extraction of Natural Products

High-Performance Green Extraction of Natural Products

Editor

Dimitris P. Makris

MDPI • Basel • Beijing • Wuhan • Barcelona • Belgrade • Manchester • Tokyo • Cluj • Tianjin

Editor
Dimitris P. Makris
Green Processes & Biorefinery Group,
Department of Food Science & Nutrition,
University of Thessaly
Greece

Editorial Office
MDPI
St. Alban-Anlage 66
4052 Basel, Switzerland

This is a reprint of articles from the Special Issue published online in the open access journal *Applied Sciences* (ISSN 2076-3417) (available at: https://www.mdpi.com/journal/applsci/special_issues/Green_Extraction_Phytochemicals).

For citation purposes, cite each article independently as indicated on the article page online and as indicated below:

LastName, A.A.; LastName, B.B.; LastName, C.C. Article Title. *Journal Name* **Year**, *Volume Number*, Page Range.

ISBN 978-3-0365-0292-2 (Hbk)
ISBN 978-3-0365-0293-9 (PDF)

Contents

About the Editor

Dimitris P. Makris was born in Xanthi (Greece), in 1974, and holds a BSc from the School of Food Technology & Nutrition of the T.E.I. of Athens (Greece, 1995), an MSc from the I.U.V.V. (Dijon, France, 1997) and a PhD in food chemistry from Imperial College (London, U.K., 2001). Since 2018, Dimitris has held the permanent position of Associate Professor in the Department of Food Science & Nutrition, School of Agricultural Sciences, University of the Thessaly. Dimitris has published over 150 research papers and book chapters, and has over than 7000 citations. Dimitris's current research interests include food waste and biomass valorization for the production of high-value-added products (natural antioxidants, food additives, etc.), through the modeling and optimization of advanced extraction technologies.

Editorial

Editorial "High-Performance Green Extraction of Natural Products"

Dimitris P. Makris

Green Processes & Biorefinery Group, Department of Food Science & Nutrition, University of Thessaly, N. Temponera Street, 43100 Karditsa, Greece; dimitrismakris@uth.gr

Received: 21 October 2020; Accepted: 28 October 2020; Published: 30 October 2020

There has been, to-date, a large number of studies pertaining to the exploitation of plant resources (herbs, botanicals, processing by-products) for the production of extracts enriched with bioactive substances. The interest has been focused on the development of efficient and cost-effective downstream processes, which aim at producing commodities on the basis of either crude or purified extracts.

Traditional extraction techniques, including percolation, decoction/infusion generation, maceration, etc., are currently being replaced by cutting-edge, sophisticated technologies with higher efficiency and selectivity, and a more eco-friendly profile. Advanced extraction methodologies based on ultrasonication, microwaves, pulsed electric fields, high voltage discharges, enzymes, pressurized liquids, supercritical fluids, deep eutectic solvents, etc., have in many instances proven to be more targeted, high-performing, straightforward, fast, sustainable, fully automated, and with relatively low capital cost.

This Special Issue addresses the concept of innovative and emerging strategies that aim at effectively implementing green technologies for the extraction of bioactive compounds from plant resources.

Loukri et al. [1] examined the extraction of caffeine and chlorogenic acids from coffee pulp, a by-product of coffee production, using aqueous solutions of β-cyclodextrin (β-CD) as a non-conventional solvent. The parameters of β-CD concentration, liquid-to-solid ratio, and temperature were evaluated based on the antiradical activity and the caffeine content, by deploying the response surface methodology. The sensory profiles of brews prepared with coffee and coffee pulp with or without cyclodextrin were studied with quantitative descriptive analysis. The brew from the by-product had fruity, botanic, sweet, and sour sensory properties, and cyclodextrin was found to be able to affect the overall taste of the brew.

Dabetić et al. [2] investigated the exploitation of deep eutectic solvents (DESs) (choline chloride: citric acid and choline chloride: glucose) as solvents for extracting valuable phenolic antioxidants from grapes. Investigation was conducted on ten grape varieties, observing seeds and skin as different matrices. Overall results support that DESs (particularly choline chloride: citric acid) were comparable to conventional solvents, and in most cases even outperformed acidified aqueous ethanol with regard to extraction efficiency and antioxidant activity. Regardless of varietal distinctions, grape seeds were found to have higher antioxidant capacity compared to grape skins, in accordance with their polyphenol concentrations.

Lakka et al. [3] developed a simple, straightforward, and green extraction methodology to effectively recover potato peel polyphenols, using hydroxypropyl β-cyclodextrin (HP-β-CD). After an initial assay to identify the optimal HP-β-CD concentration that would provide increased extraction yield, optimization based on response surface methodology enabled maximization of the extraction performance. Testing of temperatures higher than 30 °C and up to 80 °C did not favor higher yields. The extracts obtained with HP-β-CD were slightly richer in polyphenols than extracts prepared with conventional solvents, such as aqueous ethanol and methanol, displaying similar antioxidant characteristics. The major polyphenols that could be identified in the extracts were neochlorogenic, chlorogenic, caffeic, and ferulic acids. The outcome of this study demonstrated that HP-β-CD may be used as a highly effective green means of recovering potato peel polyphenols, at near-ambient temperature.

In another study, Lakka et al. [4] established a green extraction process using a novel eco-friendly natural deep eutectic solvent, composed of glycerol and nicotinamide, to produce polyphenol-enriched extracts from *Moringa oleifera* leaves. Furthermore, sample ultrasonication prior to batch stirred-tank extraction was studied to examine its usefulness as a pretreatment step. Optimization of the extraction process through the response surface methodology showed that the maximum total polyphenol yield could be achieved after a 30 min ultrasonication pretreatment, but the difference from the yield obtained from the non-pretreated sample was statistically non-significant. Extraction kinetics revealed that the activation energy for the ultrasonication-pretreated samples was more energy-demanding, a fact attributed to phenomena pertaining to washing of the readily extracted polyphenols during pretreatment. Liquid chromatography-diode array-mass spectrometry showed that ultrasonication pretreatment may have a limited positive effect on polyphenol extractability, but the overall polyphenolic profile was identical for the ultrasonication-pretreated and non-pretreated samples.

Jin et al. [5] studied the application of deep eutectic solvents (DESs) as safe and efficient extraction media that could yield maximized skin-related bioactivities from a mixture of long-lived trees: *Ginkgo biloba* L., *Cinnamomum camphora* (L.) J. Presl., and *Cryptomeria japonica* (L.f.) D. Don, native to Asia. Various DESs were synthesized from cosmetics-compatible compounds and used to prepare leaf extracts. A DES containing glycerol and xylitol yielded the highest extractability for isoquercetin, and it was selected as the optimal solvent. Then, a series of mixtures of the tree leaves were prepared according to a simplex-centroid mixture design, and their DES-extracts were tested for skin-related activities, including antioxidant, anti-tyrosinase, and anti-elastase activities. The mixture design resulted in two special cubic models and one quadratic model best fitted for describing the antioxidant and anti-elastase activities, and the anti-tyrosinase activity, respectively. Based on the established models, three different optimal formulations of the three kinds of tree leaves were suggested for maximized responses. This strategy, based on the simplex-centroid mixture design with a DES as the extraction solvent, was proposed for the development of new materials from a mixture of natural resources, suitable for the cosmetics and related fields.

Finally, in their informative review, Detsi and Skarpalezos [6] attempted to summarize the use of deep eutectic solvents in the extraction of flavonoids, one of the most important classes of plant secondary metabolites. All of the applications reviewed reported success in isolation and extraction of the target compounds: competitive, if not superior, extraction rates compared with conventional solvents; and satisfactory behavior of the extract in the latter applications (such as direct analysis, synthesis, or catalysis).

Funding: This research received no external funding.

Conflicts of Interest: The author declares no conflict of interest.

References

1. Loukri, A.; Tsitlakidou, P.; Goula, A.; Assimopoulou, A.; Kontogiannopoulos, K.; Mourtzinos, I. Green extracts from coffee pulp and their application in the development of innovative brews. *Appl. Sci.* **2020**, *10*, 6982. [CrossRef]

2. Dabetić, N.; Todorović, V.; Panić, M.; Redovniković, I.R.; Šobajić, S. Impact of deep eutectic solvents on extraction of polyphenols from grape seeds and skin. *Appl. Sci.* **2020**, *10*, 4830. [CrossRef]

3. Lakka, A.; Lalas, S.; Makris, D. Development of a low-temperature and high-performance green extraction process for the recovery of polyphenolic phytochemicals from waste potato peels using hydroxypropyl β-cyclodextrin. *Appl. Sci.* **2020**, *10*, 3611. [CrossRef]

4. Lakka, A.; Grigorakis, S.; Kaltsa, O.; Karageorgou, I.; Batra, G.; Bozinou, E.; Lalas, S.; Makris, D. The effect of ultrasonication pretreatment on the production of polyphenol-enriched extracts from *Moringa oleifera* L. (drumstick tree) using a novel bio-based deep eutectic solvent. *Appl. Sci.* **2020**, *10*, 220. [CrossRef]
5. Jin, Y.; Jung, D.; Li, K.; Park, K.; Ko, J.; Yang, M.; Lee, J. Application of deep eutectic solvents to prepare mixture extracts of three long-lived trees with maximized skin-related bioactivities. *Appl. Sci.* **2019**, *9*, 2581. [CrossRef]
6. Skarpalezos, D.; Detsi, A. Deep eutectic solvents as extraction media for valuable flavonoids from natural sources. *Appl. Sci.* **2019**, *9*, 4169. [CrossRef]

Publisher's Note: MDPI stays neutral with regard to jurisdictional claims in published maps and institutional affiliations.

 *applied sciences*

Article

Green Extracts from Coffee Pulp and Their Application in the Development of Innovative Brews

Anastasia Loukri [1], Petroula Tsitlakidou [1], Athanasia Goula [1], Andreana N. Assimopoulou [2,3], Konstantinos N. Kontogiannopoulos [2,3] and Ioannis Mourtzinos [1,*]

[1] Department of Food Science and Technology, School of Agriculture, Aristotle University of Thessaloniki, 54124 Thessaloniki, Greece; loukriap@agro.auth.gr (A.L.); ptsitlak@agro.auth.gr (P.T.); athgou@agro.auth.gr (A.G.)

[2] Organic Chemistry Laboratory, School of Chemical Engineering, Aristotle University of Thessaloniki, 54124 Thessaloniki, Greece; adreana@eng.auth.gr (A.N.A.); kkontogi@cheng.auth.gr (K.N.K.)

[3] Center of Interdisciplinary Research and Innovation of Aristotle University of Thessaloniki, Natural Products Research Centre of Excellence (NatPro-AUTH), 54124 Thessaloniki, Greece

[*] Correspondence: mourtzinos@agro.auth.gr; Tel.: +30-2310991129

Received: 21 September 2020; Accepted: 3 October 2020; Published: 6 October 2020

Abstract: Coffee pulp, a by-product of coffee production, contains valuable compounds such as caffeine and chlorogenic acid with high antiradical activity. In this study, aqueous solutions of β-cyclodextrin (β-CD) were used as a non-conventional solvent for the extraction of targeted compounds from coffee pulp. The parameters of β-CD concentration ($C_{\beta cd}$), liquid-to-solid ratio (L/S), and temperature (T) were evaluated based on the antiradical activity (A_{AR}) and the caffeine content (C_{Caf}). The optimum operational conditions were found to be $C_{\beta cd}$: 9.25 mg/mL, L/S: 30 mL/g and T: 80 °C. The sensory profiles of brews prepared with coffee and coffee pulp with or without cyclodextrin were studied with quantitative descriptive analysis. The brew from the by-product had fruity, botanic, sweet and sourness sensory properties, and cyclodextrin was found to be able to affect the overall taste of the brew.

Keywords: chlorogenic acid; caffeine; β-cyclodextrin; coffee pulp; sensory analysis; cold brew

1. Introduction

Agricultural production leads to the accumulation of agro-food wastes and by-products. These wastes can be reused as they are a source of bioactive compounds such as phenolics that can be extracted by several conventional and non-conventional techniques.

Coffee is one of the most consumed products in the word; world coffee production for 2018/19 was estimated to be about 170,937,000 bags (60 kg/bag) [1]. For the preparation of coffee, coffee grounds must be released from the coffee cherry. During this step, many parts of the cherry are thrown away. Coffee pulp is the main by-product; other by-products of coffee production include husks, silver skin, and spent waste coffee. It is estimated that around one ton of coffee pulp is produced during the manufacture of two tons of coffee [2]. Coffee pulp is rich in cellulose, hemicellulose, phenolic compounds, and caffeine [3]. Chlorogenic acid is the main phenolic compound in coffee pulp contributing to the sensory attributes of coffee [4,5]. Specifically, it is responsible for the bitter, sour, and astringency tastes. During roasting, chlorogenic acid turns into lactones and contributes to the bitter taste of the brew [6,7]. Caffeine is another significant compound in coffee, contributing 10–30% of the final bitter taste of the brew [8].

Solid–liquid extraction is the most common technique for the extraction of polyphenols. The solvent penetrates into the plant tissue and contributes to the dilution of the bioactive compounds [9,10].

The nature of solvent, the temperature, the volume of the solvent and the time of extraction can control the final yield of the process. A novel approach in extraction processes is the development of environmentally-friendly systems, such as deep eutectic solvents and solutions of cyclodextrins [9]. Cyclodextrins are cyclic oligosaccharides which are composed of six to eight glucopyranose units. Cyclodextrin has a hydrophobic cavity and a hydrophilic surface. This structure may be used as a vessel for the extraction of hydrophobic compounds that create an inclusion complex with cyclodextrin [11]. Recent studies have investigated the extraction ability of cyclodextrin solutions and reported efficiencies that were adequate, and in some cases even higher than those obtained with conventional solvents [12–14]. Moreover, cyclodextrins can be used as masking agents. Based on the characteristic molecular structure of cyclodextrins, they can capture specific compounds and prevent the interaction with taste receptors [15]. Several scientific reports have investigated the creation of complexes between cyclodextrins and the bitter compounds in coffee, such as chlorogenic acid and caffeine [16–18].

Coffee brewing is a solid–liquid extraction which affects the quality and flavor of the drink. Commercial coffee vendors have invested in cold extraction, suggesting that a cold brew coffee possesses a different sensory profile than a conventionally brewed coffee [19]. Due to the huge coffee consumption, large amounts of by-products are generated in the coffee industry. The current study suggests that by-products from the coffee industry can be utilized as potential functional ingredients. Coffee pulp valorization could lead to the development of innovative brews containing substantial amounts of functional components, such as caffeine and chlorogenic acid, which may lead to health benefits in drinkers [20], while demonstrating unique and appealing sensory properties. However, coffee pulp has not been investigated for its sensory properties.

The purpose of this study was the optimization of the extraction process of bioactives from coffee pulp using eco-friendly β-cyclodextrin solutions. The optimization process was based on a central composite design and the responses considered were the antiradical activity and the caffeine content. Moreover, the sensory profile of the brews obtained from coffee pulp was examined.

2. Materials and Methods

2.1. Chemicals and Reagents

Folin–Ciocalteu phenol reagent was obtained from Merck (Darmstradt, German). β-cyclodextrin (CD, molecular weight of ~1135), gallic acid, caffeine (99%), Trolox™, 2,2-diphenyl-picrylhydrazyl (DPPH•) stable radical were obtained from Sigma-Aldrich, Chemie GmbH (Taufkirchen, Germany). Dichloromethane (99.8%) and chlorogenic acid (Chemical Reference Standard) were supplied from Che-Lab NM (Zedelgem, Belgium) and the European Directorate for the Quality of Medicines (Starsbourg, France), respectively. All of the organic solvents used for extraction studies were of analytical grade and purchased from Sigma-Aldrich, Chemie GmbH (Taufkirchen, Germany). All UHPLC grade solvents (methanol and formic acid) were purchased from Sigma-Aldrich, Chemie GmbH (Taufkirchen, Germany) and water for HPLC was produced form a Milli-Q apparatus (Merck KGaA, Darmstadt, German).

2.2. Plant Material

Coffee pulp from the *Cutuai* (*Coffea arabica*) variety with a mean particle size of 0.24 mm was supplied by Peralta Coffees, Nicaragua. The constituents of plant material were as follows: moisture 10.36% *w/w*, protein 7.71% *w/w*, lipid 0.75% *w/w*, ash 5.23% *w/w*, carbohydrates 75.95% *w/w*, raw fiber 32.73% *w/w*. For sensory analysis, Arabica coffee seeds were bought from a local supermarket.

2.3. Experimental Design and Response Surface Methodology

A central composite design was applied to determine the effects and the optimum levels of the examined parameters. The variation of extraction yield-dependent variable was studied at different temperatures (T, 30–80 °C), cyclodextrin (CD) concentrations ($C_{\beta cd}$, 0–18.5 mg/mL), and liquid-to-solid ratio values (L/S, 13–47 mL/g). The experimental conditions were selected based on data from the literature. A three-factor, five-level central composite rotatable design (2^3 + star) was used. This design consisted of three groups of design points, including two-level factorial design points, axial or star points, and center points. Therefore, the three selected independent variables were studied at five different levels, coded as $-\alpha$, -1, 0, 1, and $+\alpha$ (Table 1). The value for alpha (1.68) was chosen to fulfill the rotatability in the design. According to the central composite design matrix, a total of 20 experiments was required (Table 2). The caffeine content (C_{Caf}) and the antiradical activity (A_{AR}) were chosen as the dependent variables.

Table 1. Experimental values and coded levels of the independent variables used for the central composite design. $C_{\beta cd}$, β-cyclodextrin concentration; L/S, liquid-to-solid ratio; T, temperature.

Independent Variable	Coded Variable Levels				
	-1.68	-1.00	0	$+1.00$	$+1.68$
$C_{\beta cd}$ (% *w/v*)	0	3.75	9.25	14.75	18.5
L/S (mL/g)	13	20	30	40	47
T (°C)	30	40	55	70	80

2.4. Extraction Process

Ground coffee pulp was mixed with an aqueous solution of β-cyclodextrin of different concentrations ($C_{\beta cd}$, 0–18.5 mg/mL), in different liquid-to-solid ratios (L/S, 13–47 mL/g), at different temperatures (T, 30–80 °C) in a stoppered glass bottle based on the experimental conditions presented in Table 2. The material was subjected to extraction under stirring at 600 rpm. In all experiments, the extracts were collected after 120 min. The time of extraction was selected based on previous experiments. Following extraction, samples were centrifuged in a bench centrifuge (Hermle Z300K, Germany) at 9000 rpm for 5 min and were separated under vacuum filtration. The clear supernatant was stored in the refrigerator until used for further analysis.

Table 2. Measured and predicted values of extracts antiradical activity (A_{AR}, µmol TRE*/g) and caffeine extraction yield (C_{Caf}, mg/g) for the individual design points.

Design Point	$C_{\beta cd}$ (mg/mL)	L/S (mL/g)	T (°C)	A_{AR} (µmol TRE/g) Measured	A_{AR} (µmol TRE/g) Predicted	C_{Caf} (mg/g) Measured	C_{Caf} (mg/g) Predicted
1	3.75	20	40	20.013	19.254	4.186	4.181
2	14.75	20	40	19.862	18.615	4.016	4.047
3	3.75	40	40	22.164	21.276	4.447	4.494
4	14.75	40	40	24.202	28.323	4.438	4.364
5	3.75	20	70	32.347	28.486	4.357	4.445
6	14.75	20	70	21.019	22.167	4.496	4.463
7	3.75	40	70	25.261	26.769	4.878	4.861
8	14.75	40	70	27.117	28.136	4.863	4.883
9	0	30	55	21.949	24.453	4.698	4.638
10	18.5	30	55	27.938	25.065	4.505	4.544
11	9.25	13	55	14.201	17.102	4.151	4.110
12	9.25	47	55	27.157	23.894	4.712	4.733
13	9.25	30	30	23.634	23.023	4.073	4.080
14	9.25	30	80	30.324	30.56	4.761	4.733
15	9.25	30	55	22.425	22.882	4.579	4.567
16	9.25	30	55	21.472	22.882	4.490	4.567
17	9.25	30	55	25.615	22.882	4.611	4.567
18	9.25	30	55	22.637	22.882	4.588	4.567
19	9.25	30	55	22.679	22.882	4.568	4.567
20	9.25	30	55	22.401	22.882	4.566	4.567

TRE*: Trolox Equivalent.

2.5. Determination of Antiradical Activity (A_{AR})

The antiradical activity was determined according to a previously described protocol [21]. Briefly, an aliquot of 0.025 mL of extract was added to 0.975 mL DPPH$^\bullet$ solution (100 µM in MeOH) and the absorbance at 515 nm was read at 0 and 30 min. Trolox™ equivalents (mM TRE) were determined from a linear regression, after plotting %ΔA_{515} of known solutions of Trolox™ against concentration, where:

$$\%\Delta A_{515} = \frac{A_{515}^{t=0} - A_{515}^{t=30}}{A_{515}^{t=0}} \times 100 \tag{1}$$

Results were expressed as µmol TRE per g of coffee pulp weight.

2.6. Determination of Caffeine Extraction Yield

Caffeine quantitation was based on a previously described protocol [22]. Briefly, 12 mL of extract was mixed with an equal amount of dichloromethane. Then, using a separatory funnel, the caffeine was extracted using dichloromethane. The caffeine extraction yield (C_{Caf}) was expressed as mg caffeine per g of coffee pulp.

2.7. Determination of Total Polyphenol Yield (Y_{TP})

The total phenolic content of the brews was determined according to a protocol [21] using the Folin–Ciocalteu methodology. Yield in total polyphenols (Y_{TP}) was expressed as mg gallic acid equivalents (GAE) per g of coffee pulp weight.

2.8. Determination of Caffeine and Chlorogenic Acid Content

The quantification of caffeine and chlorogenic acid in brews was performed by ultra-high-performance liquid chromatography (UHPLC) using an ECOM spol. s r.o., Czech Republic system (model ECS05). The system is comprised of a quaternary gradient pump (ECP2010H), a gradient box with degasser (ECB2004), a column heating/cooling oven (ECO 2080), an autosampler (ECOM Alias) and diode array detector (ECDA2800 UV-VIS PDA Detector). A Phenomenex® reversed-phase column (Synergi™ Max-RP 80 Å; 4 µm particle size, 150 × 4.6 mm) was used at 25 °C. The sample injection volume was 10 µL. Chromatographic analysis was performed using a gradient of Milli-Q water with 0.1% formic acid (solvent A) and methanol with 0.1% formic acid (solvent B), at constant flow rate of 0.5 mL/min. The gradient program was as follows: solvent A was decreased from 70% to 55% after 5 min; followed by another decrease to 35% until 15 min; while it was finally reduced to 10% at 18 min. Then, solvent A was maintained at 10% for 2 min and returned to initial conditions (70% solvent A). Detection was accomplished with the diode array detector and chromatograms were recorded at 276 nm for caffeine and 330 nm for chlorogenic acid. Identification of caffeine and chlorogenic acid was performed by comparing the retention time and the UV-Vis spectra with those of reference standards, while quantification was established with the aid of calibration curves (Equations (2) and (3), respectively). All chromatographic data were analyzed using Clarity Chromatography Software v8.2 (DataApex Ltd.).

$$\text{Caffeine Concentration} \left(\frac{mg}{mL}\right) = 0.000051 * x * area - 0.000893; \ (R^2 = 0.99952) \tag{2}$$

$$\text{Chlorogenic acid concentration} \left(\frac{mg}{mL}\right) = 0.000041 * x * area + 0.000691; \ (R^2 = 0.99999) \tag{3}$$

2.9. Sensory Analysis

2.9.1. Preparation of Brews

A set of four samples was prepared for the sensory evaluation. The applied experimental conditions are displayed in Table 3 for design point 13, as these conditions were closest to the conditions of the cold brewing technique. All brews were prepared by using a liquid-to-solid ratio of 30 mL/g, an extraction time of 2 h, and a temperature of 30 °C. Two of the four samples were from Arabica coffee seeds (AQC) while the other two samples (AQCW) were made by using coffee pulp as the main extractable raw material. In each subset of samples, β-cyclodextrin was dissolved in water at a concentration of 9.25 mg/mL in one of the two samples prior to the start of the coffee extraction (AQC/CD and AQCW/CD). Detailed sample formulation is presented in Table 3. Subsequently, the extracts were collected by filtration and stored in a refrigerator until the sensory session. For each panelist, a quantity of 20 mL was served at room temperature in a plastic cup with a lid. The samples were randomly presented in four hourly sessions on three consecutive days.

Table 3. Experimental conditions for the samples used in the sensory evaluation. t, time.

Sample Name	Raw Materials	$C_{\beta cd}$ (mg/mL)	L/S (mL/g)	T (°C)	t (h)
AQC	Arabica coffee beans	0			
AQC/CD	Arabica coffee beans	9.25			
AQCW	Arabica coffee pulp	0	30	30	2
AQCW/CD	Arabica coffee pulp	9.25			

AQC: aqueous extract from coffee seeds, AQC/CD: β-cyclodextrin extract from coffee seeds, AQCW: aqueous extract from coffee pulp, AQCW/CD: β-cyclodextrin extract from coffee pulp.

2.9.2. Quantitative Descriptive Analysis

Nine trained panelists from the Department of Food Science and Technology (Aristotle University of Thessaloniki, Greece), who had already participated in several trained panel studies for other food products and with at least 1 year of experience in sensory evaluation, developed a consensus vocabulary of 11 descriptors using a Quantitative Descriptive Analysis approach. The attributes were categorized under the modalities of aroma and taste and reference standards were used when required. For a better explanation of the developed attributes, the reference standards are presented in Table 4. The terms botanic, fruity, earthy, and roasted were used to describe the aroma characteristics, whereas the words bitter, sweet, sourness, sour-roasted, botanic, earthy, and astringency were used to characterize the taste of the samples.

Table 4. Developed attributes and references standards in the sensory evaluation.

Modality	Descriptor	Definition
Odor	Botanic	Characteristic odor associated with typical dried black tea notes
	Fruity	Overall odor associated with floral, sweet, ripe fruits and characteristic odor of coffee pulp
	Earthy	Odor associated with bread and wet soil
	Roasted	Characteristic odor of over-roasted hazelnuts (220 °C/10 minutes)
Taste	Bitter	The fundamental sensation associated with caffeic acid
	Sweet	The fundamental sensation associated with sucrose
	Sourness	Taste associated with sour/fermented-like aromatics
	Sour-roasted like	Characteristic acidic, sharp and pungent taste associated with excessively roasted coffee beans
	Botanic	Characteristic taste of black tea infusion
	Earthy	Characteristic taste associated with bread crust
Tactile sensation	Astringent	A dry penetrating sensation in the nasal cavity

Attributes were scored using unstructured line scales (0–100) and panelists were seated in booths under appropriate environmental conditions (in accordance with International Organization for

Standardization recommendation (ISO 8589:2010). Panelists individually rated samples in duplicate on two separate examinations. Samples were presented monadically according to a balanced design, labeled with arbitrary three-digit codes in opaque white plastic cups and maintained at ambient temperature before serving for evaluation. Bottled water and unsalted crackers were provided as palate cleanser between samples.

2.10. Statistical Analysis

The response surface methodology (RSM) data were analyzed using the statistical software MINITAB (release 13.32). To identify the significance of the effects and interactions between them, analysis of variance (ANOVA) was performed for each p-value less than 0.05, which was considered to be statistically significant. Regression analysis was used to fit a full second order polynomial to the data of the response variables. To evaluate the goodness of fit of each model, two criteria were used: the coefficient of determination, R^2, which is the relative variance explained by the model with respect to the total variance.

Sensory data were analyzed using the SENPAQ software (Qi Statistics, Ruscombe, UK). A two-way model analysis of variance (ANOVA) was performed; panelists were treated as random effects and samples and replicates as fixed effects. Multiple pairwise comparisons were conducted by Tukey's method and a significant difference was determined at an alpha risk of 5% ($p \leq 0.05$). All determinations were carried out in triplicate. The values obtained were averaged.

3. Results and Discussion

3.1. Extraction Yield

Figure 1 presents the effect of the process variables on the responses A_{AR} and C_{Caf}. In the case of A_{AR}, it can be observed that an increase of $C_{\beta cd}$ generally leads to increased values of A_{AR}. Similar studies on plant extracts rich in phenolics showed that the extract antiradical activity increased with the concentration of β-cyclodextrin in the solution [23,24], since the existence of β-CD can lead to inclusion complex formation with phenolic compounds. According to [9], the protection of phenolic compounds with β-CD can contribute to increased values of A_{AR}. However, the β-CD concentration does not seem to affect the concentration of caffeine in the extract and it can be observed that the highest C_{Caf} was obtained in the absence of cyclodextrin. The inclusion complex formation of caffeine has already been studied [25] and the usage of cyclodextrin has been proposed for the decaffeination process [26]. In complicated extracts, it is difficult to figure out which specific compounds form complexes [13]. It seems that the presence of cyclodextrin preferably boosted the extraction of more hydrophobic polyphenol molecules than caffeine [27].

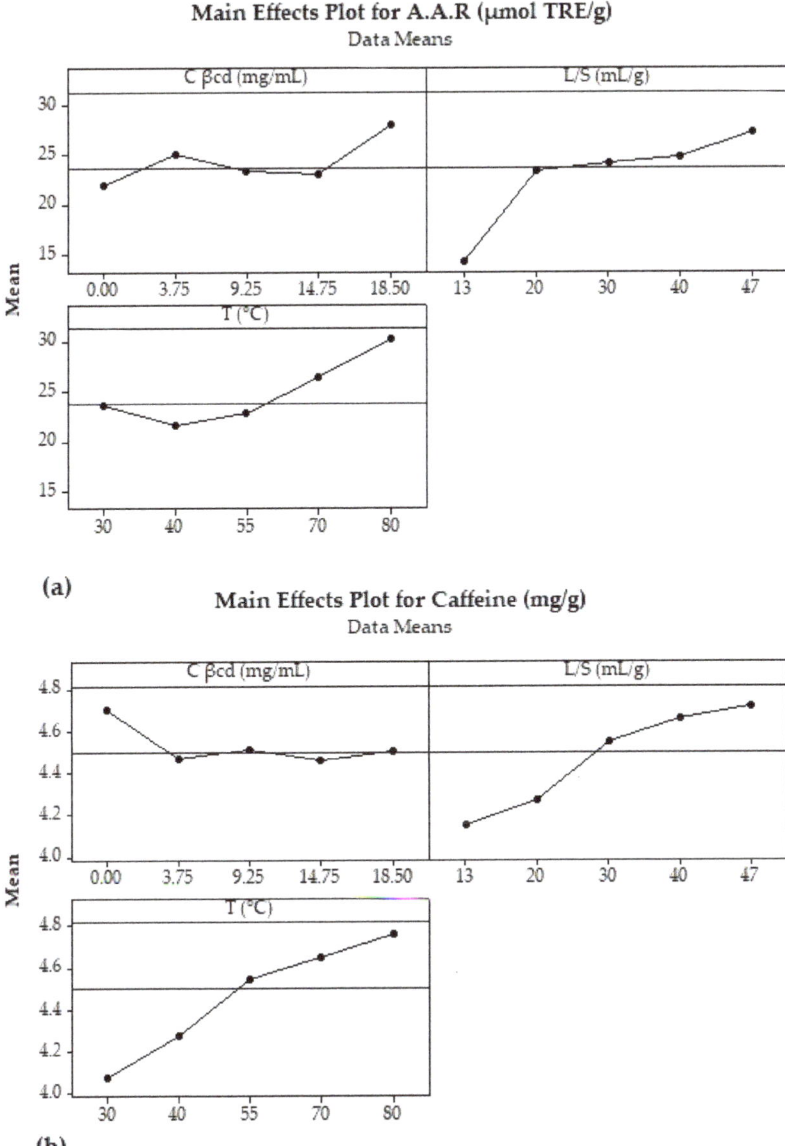

Figure 1. Main effects plots presenting the effect of β-cyclodextrin concentration ($C_{\beta cd}$, mg/mL), liquid-to-solid ratio (L/S, mL/g), and temperature (T, °C) (**a**) on extract antiradical activity (A_{AR}, µmol TRE/g) and (**b**) on caffeine extraction yield (C_{Caf}).

The liquid-to-solid ratio (L/S) has a great impact on the examined responses. The extraction yield increased with increasing ratio. A similar trend was reported by many researchers [12,28,29]. A higher ratio results in a larger concentration gradient during diffusion from the solid into the solution and in excessive swelling of the plant material, increasing the contact surface area between the material and the solvent [30].

The positive effect of extraction temperature on A_{AR} and C_{Caf} is presented in Figure 1. It is widely recognized that temperature enhances mass transfer by improving the extraction rate; a fact that can be attributed to the effect of temperature on the vapor pressure, surface tension, and viscosity of the solvent. Moreover, the increase in the yield may be associated with the increased ease with which solvent diffuses into cells and the enhancement of desorption and solubility at high temperatures [31]. Previous studies have reported a decrease of extract antiradical activity with temperature due to the possible degradation of polyphenolic compounds [32].

The regression coefficients were calculated and the data was fitted to second-order polynomial equations:

$$A_{AR}(\mu mol\ TRE/g) = 12.32 - 0.474 * C_{\beta\ cd} + 0.714 * L/s - 0.191 * T + 0.022 * C^2_{\beta\ cd} - 0.008 * (L/s)^2$$
$$+ 0.006 * T^2 + 0.035 * C_{\beta\ cd} * L/s - 0.017 * C_{\beta\ cd} * T - 0.006 * L/s * T \tag{4}$$

$$C_{Caf}\ (mg/g) = 2.660 - 0.036 * C_{\beta\ cd} + 0.039 * L/s + 0.032 * T + 0.280 * 10^{-3} * C^2_{\beta\ cd} - 0.500 * 10^{-3} * (L/s)^2$$
$$-0.260 * T^2 + 0.020 * 10^{-3} * C_{\beta\ cd} * L/s + 0.460 * 10^{-3} * C_{\beta\ cd} * T + 0.170 * 10^{-3} * L/s * T \tag{5}$$

The coefficient of determination, R^2, for A_{AR} and C_{Caf} was 0.729 and 0.969, respectively, indicating that 72.9% and 96.9% of the total variability in the responses could be explained by the specific models. According to the regression analysis, the temperature, the liquid-to-solid ratio, and their quadrates significantly influence the extraction yield ($p < 0.05$).

The measured and the predicted values of the responses are analytically depicted in Table 2. The extraction conditions of 9.25 mg/mL ($C_{\beta cd}$), 30 mL/g (L/S), and 80 °C (T) are the optimum conditions. Under these conditions, the predicted values of antiradical activity and caffeine content were 30.56 µmol TRE/g dry matter and 4.733 mg/g, respectively, whereas the observed experimental values were 30.324 µmol TRE/g dry matter and 4.761 mg/g, respectively.

3.2. Sensory Analysis

A consensus vocabulary of eleven attributes was developed to describe the samples, under the modalities of odor, taste, and mouthfeel, while using reference standards where required. Similar sensory descriptors were generated by other trained panels to characterize brewed coffee beverages [33]. All sensory descriptors differ significantly between samples ($p < 0.05$, Table 5). To the best of the authors' knowledge, this is the first time that coffee pulp has been used scientifically for cold brewed coffee drink preparation and the developed lexicon is unprecedented. Cold brewing is an innovative technique and several researchers have worked on its sensory evaluation. Generally, cold brews are characterized by a sweeter and sour taste as compared to the product of the classical technique. Hence, the purpose of the sensory profiling was to investigate the sensory quality of an alternative coffee drink/beverage formulation and provide a measure of the alterations in coffee drink descriptors occurring when the raw material used is changed.

The sensory profiles of the different brews are presented in Figure 2. Brews prepared from coffee grounds (AQC and AQC/CD) were compared to the brews obtained from coffee pulp (AQCW and AQCW/CD). The panelists scored the latter samples as significantly more botanic and fruitier in odor and more sweet, sour, and botanic in taste. The earthy odor and taste intensities were significantly higher for the coffee bean brew samples as compared to the other two samples. In addition, the AQC and AQC/CD samples had significantly higher scores in bitter and sour-roasted taste than their counterparts. The results of this study show that the sensory characteristics of the cold brew coffee bean samples resembled the usual organoleptic properties of coffee samples prepared with the hot brewing technique. Attributes such as earthy, roasted, and fruity aromas, bitter and sour tastes, and astringency had been previously used to evaluate hot brewed coffee [33]; however, the perceived intensities are not similar. On the contrary, the sensory attributes of the AQC and AQC/CD samples possessed similar characteristics to tea beverages, since they were characterized by high intensities of botanic and fruity aromas, and sweet, sour, and botanic taste sensations.

Table 5. Mean intensity scores of aroma and taste attributes for the different brews.

Characteristics **	Attributes	Samples			
		AQC ***	AQC/CD	AQCW	AQCW/CD
O	Botanic	7.8 [b]	6.4 [b]	54.6 [a]	52.1 [a]
O	Fruity	5.8 [b]	1.8 [b]	49.5 [a]	33.2 [a]
O	Earthy	30.7 [a]	31.9 [a]	11.9 [b]	13.8 [b]
O	Roasted	61.6 [a]	60.5 [a]	1.1 [b]	1.6 [b]
T	Bitter	56.4 [a]	35.9 [b]	3.1 [c]	2.3 [c]
T	Sweet	6.2 [c]	5.9 [c]	23.5 [b]	38.5 [a]
T	Sourness	1.1 [b]	2.1 [b]	49.4 [a]	43.6 [a]
T	Sour-roasted	26.3 [a]	37.5 [a]	2.0 [b]	1.4 [b]
T	Botanic	4.9 [b]	4.5 [b]	42.5 [a]	38.6 [a]
T	Earthy	33.1 [a]	31.1 [a]	8.7 [b]	9.5 [b]
MF	Astrigency	21.9 [ab]	24.6 [a]	17.7 [ab]	10.1 [b]

abc: different letters represent statistically significant differences, $p < 0.05$ (Tukey's method). **: O for odor, T for taste, MF for mouthfeel. ***: AQC: aqueous extract from coffee seeds, AQC/CD: β-cyclodextrin extract from coffee seeds, AQCW: aqueous extract from coffee pulp, AQCW/CD: β-cyclodextrin extract from coffee pulp.

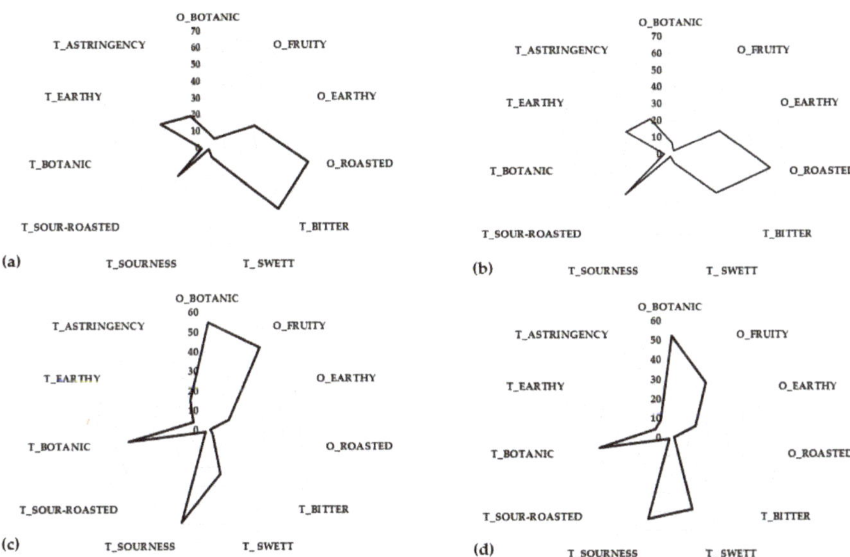

Figure 2. Spider graph for the sensory attributes of the different brews. (**a**) AQC: aqueous extract from coffee seeds, (**b**) AQC/CD: β-cyclodextrin extract from coffee seeds, (**c**) AQCW: aqueous extract from coffee pulp, (**d**) AQCW/CD: β-cyclodextrin extract from coffee pulp.

As far as the effect of β-CD is concerned, it was found that the AQC sample had a less bitter taste intensity as compared to the AQC/CD sample. Based on the fact that the content of chlorogenic acid and caffeine was the same in both brews (Table 6), it seems that the selective inclusion formation of the extracted ingredients affects the sensory characteristics of the brews. The same behavior has been observed in tea infusion, in which cyclodextrin appeared selective in the extraction of different compounds from tea [34]. These results give a new insight in the contribution of cyclodextrin in taste differentiation. Cyclodextrin has been used for taste masking in different products. In bitter gourd juice, cyclodextrin contributes to the reduction of the bitter taste [35] and this effect was associated with the fact that the molecules responsible for bitter sensations, caffeine and chlorogenic acid, partially form stable inclusion complexes with β-CD [36], resulting in masking of the bitter sensation by more than 30%. The mechanism of complex formation between β-CD and chlorogenic and caffeic acids in aqueous solutions was previously investigated, suggesting that catechol hydroxyl groups are trapped inside the β-CD cavity [17]. Another effect of β-CD addition is related to the intensity of astringency. More specifically, the coffee pulp brewed sample with β-CD was characterized by a less astringent mouthfeel as compared to the other coffee bean brewed sample.

Table 6. Total phenolics, caffeine, and chlorogenic acid content in the different brews.

	AQC *	AQC/CD	AQCW	AQCW/CD
Total phenolics (mg GAE/mL)	558.96 ± 4.18	684.45 ± 9.02	278.99 ± 4.51	198.30 ± 2.95
Caffeine (mg/mL)	0.4211	0.4257	0.1357	0.1388
Chlorogenic acid (mg/mL)	0.1065	0.1086	0.152	0.151

* AQC: aqueous extract from coffee seeds, AQC/CD: β-cyclodextrin extract from coffee seeds, AQCW: aqueous extract from coffee pulp, AQCW/CD: β-cyclodextrin extract from coffee pulp.

β-CD can be considered as a slightly sweet substance, showing a recognition threshold for sweetness in water around 0.5% [36]; hence, its sweetness cannot be ignored. This study showed two results concerning sweetness: there was no perceived variation in sweet taste between the coffee bean brewed samples, whereas sweetness was more pronounced in AQCW/CD than in the sample prepared by brewing coffee pulp with water. This observation could be attributed to the phenomenon of mixture intensity suppression; in the case of coffee brew samples, intensity was at such high levels that it suppressed the perception of sweet taste, whereas the coffee pulp brewed samples scored very low in bitterness, "allowing" the perception of different intensities in sweetness.

Earlier studies utilized coffee by-products as a source of antioxidants and dietary fibers in bakery products, such as biscuits [37] and muffins [38], investigating consumer acceptance and the sensory quality of the novel products. There is currently very little research published on the development of functional beverages from coffee waste. Coffee silver skin was brewed and used as an ingredient in a beverage aiming to reduce body fat accumulation [39]. The study concluded that the acceptance level of the beverages made with coffee silver skin was about 95%. However, the effect of coffee by-product use on the sensory profile of the functional beverages was not investigated. To the best of our knowledge, the current study is the first systematic approach in determining variations in sensory attributes of cold brew coffees made from coffee seeds or coffee waste with or without β-CD addition. The brewed coffee waste samples demonstrated sensory properties similar to beverages based on tea brew, with highly perceived intensities of fruity and botanic aroma characteristics and sweet, sour, and botanic taste characteristics. This novel beverage prepared by utilizing a coffee by-product could be widely accepted for its sensory quality and consumed by non-coffee drinkers as well.

4. Conclusions

The present study showed, for the first time, that aqueous solutions of β-cyclodextrin could be very effective solvent systems regarding the extraction of phytochemicals from coffee pulp. The experimental design based on a response surface methodology optimized the conditions, enabling the production of extracts with enhanced antioxidant and caffeine content. The applied extraction method with cyclodextrin resulted in innovative brews from coffee and coffee by-products with modified sensory attributes.

Author Contributions: Conceptualization, I.M.; methodology, A.L., P.T., A.G. and I.M.; formal analysis, A.L., P.T. and A.G.; investigation, A.L. and K.N.K.; resources, I.M.; writing—original draft preparation, A.L.; writing—review and editing, A.G., A.N.A. and I.M.; supervision, I.M.; project administration, I.M. All authors have read and agreed to the published version of the manuscript.

Funding: This research received no external funding.

Acknowledgments: Andrena N. Assimopoulou and Konstantinos N. Kontogiannopoulos would like to acknowledge the support of this work from the project "Upgrading the Plant Capital (PlantUp)" (MIS 5002803), which is implemented under the Action "Reinforcement of the Research and Innovation Infrastructure," funded by the Operational Programme "Competitiveness, Entrepreneurship and Innovation" (NSRF 2014–2020) and co-financed by Greece and the European Union (European Regional Development Fund).

Conflicts of Interest: The authors declare no conflict of interest.

References

1. International Coffee Organization. Total Production. Available online: http://www.ico.org/historical/1990onwards/PDF/1a-total-production.pdf (accessed on 12 May 2020).
2. Alves, R.C.; Rodrigues, F.; Antónia Nunes, M.; Vinha, A.F.; Oliveira, M.B.P.P. State of the Art in Coffee Processing By-Products. In *Handbook of Coffee Processing By-Products*; Galanakis, C.M., Ed.; Academic Press: Cambridge, MA, USA, 2017; pp. 1–26. [CrossRef]
3. Esquivel, P.; Jiménez, V.M. Functional Properties of Coffee and Coffee By-Products. *Food Res. Int.* **2012**, *46*, 488–495. [CrossRef]
4. Heeger, A.; Kosińska-Cagnazzo, A.; Cantergiani, E.; Andlauer, W. Bioactives of Coffee Cherry Pulp and Its Utilisation for Production of Cascara Beverage. *Food Chem.* **2017**, *221*, 969–975. [CrossRef]
5. de Souza Gois Barbosa, M.; dos Santos Scholz, M.B.; Kitzberger, C.S.G.; de Toledo Benassi, M. Correlation between the Composition of Green Arabica Coffee Beans and the Sensory Quality of Coffee Brews. *Food Chem.* **2019**, *292*, 275–280. [CrossRef]
6. Farah, A.; De Paulis, T.; Trugo, L.C.; Martin, P.R. Effect of Roasting on the Formation of Chlorogenic Acid Lactones in Coffee. *J. Agric. Food Chem.* **2005**, *53*, 1505–1513. [CrossRef]
7. Seninde, D.R.; Chambers, E. Coffee Flavor: A Review. *Beverages* **2020**, *6*, 44. [CrossRef]
8. Frank, O.; Blumberg, S.; Kunert, C.; Zehentbauer, G.; Hofmann, T. Structure Determination and Sensory Analysis of Bitter-Tasting 4-Vinylcatechol Oligomers and Their Identification in Roasted Coffee by Means of LC-MS/MS. *J. Agric. Food Chem.* **2007**, *55*, 1945–1954. [CrossRef]
9. Cai, R.; Yuan, Y.; Cui, L.; Wang, Z.; Yue, T. Cyclodextrin-Assisted Extraction of Phenolic Compounds: Current Research and Future Prospects. *Trends Food Sci. Technol.* **2018**, *79*, 19–27. [CrossRef]
10. Goula, A.M. Ultrasound-Assisted Extraction of Pomegranate Seed Oil–Kinetic Modeling. *J. Food Eng.* **2013**, *117*, 492–498. [CrossRef]
11. Kurkov, S.V.; Loftsson, T. Cyclodextrins. *Int. J. Pharm.* **2013**, *453*, 167–180. [CrossRef]
12. Mourtzinos, I.; Menexis, N.; Iakovidis, D.; Makris, D.; Goula, A. A Green Extraction Process to Recover Polyphenols from Byproducts of Hemp Oil Processing. *Recycling* **2018**, *3*, 15. [CrossRef]
13. Diamanti, A.C.; Igoumenidis, P.E.; Mourtzinos, I.; Yannakopoulou, K.; Karathanos, V.T. Green Extraction of Polyphenols from Whole Pomegranate Fruit Using Cyclodextrins. *Food Chem.* **2017**, *214*, 61–66. [CrossRef] [PubMed]
14. Rajha, H.N.; Chacar, S.; Afif, C.; Vorobiev, E.; Louka, N.; Maroun, R.G. β-Cyclodextrin-Assisted Extraction of Polyphenols from Vine Shoot Cultivars. *J. Agric. Food Chem.* **2015**, *63*, 3387–3393. [CrossRef] [PubMed]
15. Astray, G.; Gonzalez-Barreiro, C.; Mejuto, J.C.; Rial-Otero, R.; Simal-Gándara, J. A Review on the Use of Cyclodextrins in Foods. *Food Hydrocoll.* **2009**, *23*, 1631–1640. [CrossRef]
16. Prabu, S.; Swaminathan, M.; Sivakumar, K.; Rajamohan, R. Preparation, Characterization and Molecular Modeling Studies of the Inclusion Complex of Caffeine with Beta-Cyclodextrin. *J. Mol. Struct.* **2015**, *1099*, 616–624. [CrossRef]
17. Górnas, P.; Neunert, G.; Baczyński, K.; Polewski, K. Beta-Cyclodextrin Complexes with Chlorogenic and Caffeic Acids from Coffee Brew: Spectroscopic, Thermodynamic and Molecular Modelling Study. *Food Chem.* **2009**, *114*, 190–196. [CrossRef]
18. Aree, T. Understanding Structures and Thermodynamics of β-Cyclodextrin Encapsulation of Chlorogenic, Caffeic and Quinic Acids: Implications for Enriching Antioxidant Capacity and Masking Bitterness in Coffee. *Food Chem.* **2019**, *293*, 550–560. [CrossRef] [PubMed]
19. Cordoba, N.; Fernandez-Alduenda, M.; Moreno, F.L.; Ruiz, Y. Coffee Extraction: A Review of Parameters and Their Influence on the Physicochemical Characteristics and Flavour of Coffee Brews. *Trends Food Sci. Technol.* **2020**, *96*, 45–60. [CrossRef]
20. Sato, Y.; Itagaki, S.; Kurokawa, T.; Ogura, J.; Kobayashi, M.; Hirano, T.; Sugawara, M.; Iseki, K. In Vitro and in Vivo Antioxidant Properties of Chlorogenic Acid and Caffeic Acid. *Int. J. Pharm.* **2011**, *403*, 136–138. [CrossRef]

21. Arnous, A.; Makris, D.P.; Kefalas, P. Correlation of Pigment and Flavanol Content with Antioxidant Properties in Selected Aged Regional Wines from Greece. *J. Food Compos. Anal.* **2002**, *15*, 655–665. [CrossRef]

22. Belay, A.; Ture, K.; Redi, M.; Asfaw, A. Measurement of Caffeine in Coffee Beans with UV/Vis Spectrometer. *Food Chem.* **2008**, *108*, 310–315. [CrossRef]

23. Korompokis, K.; Igoumenidis, P.E.; Mourtzinos, I.; Karathanos, V.T. Green Extraction and Simultaneous Inclusion Complex Formation of Sideritis Scardica Polyphenols. *Int. Food Res. J.* **2017**, *24*, 1233–1238.

24. Tutunchi, P.; Roufegarinejad, L.; Hamishehkar, H.; Alizadeh, A. Extraction of Red Beet Extract with β-Cyclodextrin-Enhanced Ultrasound Assisted Extraction: A Strategy for Enhancing the Extraction Efficacy of Bioactive Compounds and Their Stability in Food Models. *Food Chem.* **2019**, *297*, 124994. [CrossRef] [PubMed]

25. Santos, C.I.A.V.; Teijeiro, C.; Ribeiro, A.C.F.; Rodrigues, D.F.S.L.; Romero, C.M.; Esteso, M.A. Drug Delivery Systems: Study of Inclusion Complex Formation for Ternary Caffeine–β-Cyclodextrin–Water Mixtures from Apparent Molar Volume Values at 298.15 K and 310.15 K. *J. Mol. Liq.* **2016**, *223*, 209–216. [CrossRef]

26. Yu, E.C. Novel Decaffeination Process Using Cyclodextrins. *Appl. Microbiol. Biotechnol.* **1988**, *28*, 546–552. [CrossRef]

27. Rescifina, A.; Chiacchio, U.; Iannazzo, D.; Piperno, A.; Romeo, G. β-Cyclodextrin and Caffeine Complexes with Natural Polyphenols from Olive and Olive Oils: NMR, Thermodynamic, and Molecular Modeling Studies. *J. Agric. Food Chem.* **2010**, *58*, 11876–11882. [CrossRef]

28. Wang, L.; Zhou, Y.; Wang, Y.; Qin, Y.; Liu, B.; Bai, M. Two Green Approaches for Extraction of Dihydromyricetin from Chinese Vine Tea Using β-Cyclodextrin-Based and Ionic Liquid-Based Ultrasonic-Assisted Extraction Methods. *Food Bioprod. Process.* **2019**, *116*, 1–9. [CrossRef]

29. Radojković, M.; Zeković, Z.; Sudar, R.; Jokić, S.; Cvetanović, A. Optimization of Solid-Liquid Extraction of Antioxidants and Saccharides from Black Mulberry Fruit by Response Surface Methodology. *J. Food Nutr. Res.* **2012**, *50*, 167–176.

30. Drevelegka, I.; Goula, A.M. Recovery of Grape Pomace Phenolic Compounds through Optimized Extraction and Adsorption Processes. *Chem. Eng. Process. Process Intensif.* **2020**, *149*, 107845. [CrossRef]

31. Goula, A.M.; Thymiatis, K.; Kaderides, K. Valorization of Grape Pomace: Drying Behavior and Ultrasound Extraction of Phenolics. *Food Bioprod. Process.* **2016**, *100*, 132–144. [CrossRef]

32. Favre, L.C.; dos Santos, C.; López-Fernández, M.P.; Mazzobre, M.F.; del Pilar Buera, M. Optimization of β-Cyclodextrin-Based Extraction of Antioxidant and Anti-Browning Activities from Thyme Leaves by Response Surface Methodology. *Food Chem.* **2018**, *265*, 86–95. [CrossRef]

33. Chapko, M.J.; Seo, H.-S. Characterizing Product Temperature-Dependent Sensory Perception of Brewed Coffee Beverages: Descriptive Sensory Analysis. *Food Res. Int.* **2019**, *121*, 612–621. [CrossRef] [PubMed]

34. Dai, Q.; Liu, S.; Jin, H.; Jiang, Y.; Xia, T. Effects of Additive β-Cyclodextrin on the Performances of Green Tea Infusion. *J. Chem.* **2019**, *2019*, 1–7. [CrossRef]

35. Deshaware, S.; Gupta, S.; Singhal, R.S.; Joshi, M.; Variyar, P.S. Debittering of Bitter Gourd Juice Using β-Cyclodextrin: Mechanism and Effect on Antidiabetic Potential. *Food Chem.* **2018**, *262*, 78–85. [CrossRef] [PubMed]

36. Szejtli, J.; Szente, L. Elimination of Bitter, Disgusting Tastes of Drugs and Foods by Cyclodextrins. *Eur. J. Pharm. Biopharm.* **2005**, *61*, 115–125. [CrossRef] [PubMed]

37. Martinez-Saez, N.; García, A.T.; Pérez, I.D.; Rebollo-Hernanz, M.; Mesías, M.; Morales, F.J.; Martín-Cabrejas, M.A.; del Castillo, M.D. Use of Spent Coffee Grounds as Food Ingredient in Bakery Products. *Food Chem.* **2017**, *216*, 114–122. [CrossRef] [PubMed]

38. Severini, C.; Caporizzi, R.; Fiore, A.G.; Ricci, I.; Onur, O.M.; Derossi, A. Reuse of Spent Espresso Coffee as Sustainable Source of Fibre and Antioxidants. A Map on Functional, Microstructure and Sensory Effects of Novel Enriched Muffins. *LWT-Food Sci. Technol.* **2020**, *119*, 108877. [CrossRef]

39. Martinez-Saez, N.; Ullate, M.; Martin-Cabrejas, M.A.; Martorell, P.; Genovés, S.; Ramon, D.; del Castillo, M.D. A Novel Antioxidant Beverage for Body Weight Control Based on Coffee Silverskin. *Food Chem.* **2014**, *150*, 227–234. [CrossRef]

applied sciences

Article

Impact of Deep Eutectic Solvents on Extraction of Polyphenols from Grape Seeds and Skin

Nevena Dabetić [1], Vanja Todorović [1], Manuela Panić [2], Ivana Radojčić Redovniković [2,*] and Sladjana Šobajić [1]

[1] Faculty of Pharmacy, University of Belgrade, Vojvode Stepe 450, 11221 Belgrade, Serbia;
nevenad@pharmacy.bg.ac.rs (N.D.); vanja.todorovic@pharmacy.bg.ac.rs (V.T.);
sladjana.sobajic@pharmacy.bg.ac.rs (S.S.)

[2] Faculty of Food Technology and Biotechnology, University of Zagreb, Pierottijeva 6, 10000 Zagreb, Croatia;
mpanic@pbf.hr

* Correspondence: irredovnikovic@pbf.hr

Received: 5 June 2020; Accepted: 11 July 2020; Published: 14 July 2020

Featured Application: The interest for green and sustainable utilization of industry by-products, such as grape pomace, is increasing rapidly, and is in line with the recognition of the necessity to protect the environment. Our investigation, mainly focused on natural deep eutectic solvents (NADES), could contribute to further exploitation of these solvents for extraction of biologically active compounds from different plants. Finally, the present study provides profound understanding of antioxidant activity and polyphenol composition of grape seeds and skin, highlighting the quality of some insufficiently investigated Serbian autochthonous varieties and suggesting their broaden use.

Abstract: In the past few years, research efforts have focused on plant exploitation for deriving some valuable compounds. Extraction has been usually performed using petrochemical and volatile organic solvents, but nowadays, increased recognition of environmental pollution has prompted the utilization of green solvents as alternatives. Therefore, the aim of the present study was to exploit deep eutectic solvents (DES) (choline chloride: citric acid and choline chloride: glucose) as solvents for extracting valuable phenolic antioxidants from grapes. Investigation was conducted on ten grape varieties, observing seeds and skin as different matrix. Total polyphenol content (TPC) was determined by *Folin-Ciocalteu* spectrophotometric microassay. Antioxidant activity was investigated using four different tests and results were combined in a unique Antioxidant Composite Index (ACI) to reveal comprehensive information about this biological activity. Polyphenol compounds were identified and quantified with the aim of HPLC-diode array detector (DAD). Overall results support that DES (particularly choline chloride: citric acid) were comparable to conventional solvent, and in most cases even outperformed acidified aqueous ethanol (concerning extraction efficiency and antioxidant activity). Regardless of varietal distinctions, grape seeds have higher antioxidant capacity compared to grape skin, and such findings are according to their phenol compound concentrations.

Keywords: grape seed; grape skin; green extraction; deep eutectic solvents; antioxidant activity; polyphenols

---

1. Introduction

Grape (*Vitis Vinifera* L. ssp sativa) is generally cultivated in moderate-warm climate zones, and it is of worldwide interest for nutritional and medical purposes [1]. This fruit, as one of the main natural dietary sources of polyphenols, is associated with numerous health benefits. It is important to emphasize that phenolic composition is affected by the grape variety, degree of ripeness, as well as agronomic and environmental conditions [2]. Moreover, there is a significant difference between parts of fruits (seeds, skin, and stems) considering not only the number of phenolic compounds, but also their concentrations [3]. There are three main classes of polyphenols abounding in grapes: phenolic acids, flavonoids, and stilbenes, each with their own family members. Numerous studies suggest that polyphenols work in more than one way, expressing antioxidant activity and influencing cell communications that affect important biological processes. In particular, grape polyphenols could contribute to a healthy heart by promoting the relaxation of blood vessels to help maintain healthy blood flow and function [4]. Extensive information is available on grape antioxidant activity, concerning the prevention of the degenerative pathophysiological state developed in healthy adults [5].

Approximately 75 million tons of grapes are produced every year worldwide, whereby almost 80% of this quantity is utilized in wine production [6]. Progressively increasing amounts of wine waste represents a serious environmental pollution problem [7,8]. Amid wine generation, most of the valuable health promoting compounds from grape berries are extricated into juice or wine, but a noteworthy sum remains caught within the pomace (skin and seeds). An increasing body of evidence has recently appeared that the revalorization of wine by-products is conceivable. Grape pomace, such as value-added products containing plenty of active compounds, have been proposed for enrichment of food, pharmaceutic, or even cosmetic items [9].

Extraction of valuable and health-promoting constituents was usually performed using petrochemical and volatile organic solvents. Most of these solvents are of serious concern, since they are flammable, low biodegradable, toxic, and volatile. Along with the growing recognition of the human impact on the environment, green extraction has become an optimal and sustainable way for raw material utilization [10]. Nowadays, several types of solvents, including natural deep eutectic solvents (NADES), have been suggested as an alternative for conventional solvents. NADES are prepared by mixing quaternary ammonium salts (e.g., choline chloride) and naturally derived hydrogen bond donors (e.g., sugars, alcohols, amines) [11]. Commonly, NADES components are non-toxic, cheap, and readily available [12]. Due to their unique physicochemical properties, such as adjustable viscosity, wide polarity range, and high solubilization strength for a broad range of compounds variety, NADES have a great potential for different health related purposes [13]. They were confirmed as excellent solvents for sustainable and environmentally friendly extraction. Choline chloride based natural deep eutectic solvents were demonstrated to have high stabilizing ability for phenolic compounds, which can be correlated with the strong hydrogen bonding interactions between solutes and solvent molecules [14]. Besides the extraction, they have been used in enzyme stability and enzymatic reactions [15–17]. Furthermore, the literature indicates that natural deep eutectic solvents could be designed with specific biological activity. They were reported to play an important role in enhancing antioxidative activities of plant extracts by possessing this activity itself [12].

Serbia has a long viticulture tradition (it is one of the oldest grapevine growing areas in Europe) due to its favorable geological position. Concerning the rapid growth of the wine industry, there was a need to perform comprehensive research on biologically active compounds of the grape that would include different varieties cultivated in this area. To the best of our knowledge, most of the studies explored the composition and bioactivity of seeds and/or skins obtained from international varieties [18–20]. Investigations of typical Serbian grapes have been undertaken using only a few autochthonous varieties [21–24]. Keeping in mind all of the mentioned above, the main aim of the present study was to determine total phenolic content and antioxidant capacity of grape seed and skin extracts obtained from ten grape varieties (both international and autochthonous varieties were included). Moreover, the present study was designed as an opportunity to broaden previous investigations of extraction efficiency of NADES [10,25,26] Thus, the complementary aim was to exploit deep eutectic solvents (choline chloride: glucose and choline chloride: citric acid) as solvents for extracting valuable phenolic compounds from the grape. Overall results showed that acid based NADES outperformed acidified aqueous ethanol, concerning both seeds and skin. There were no significant differences between widely used international and autochthonous varieties. Additionally, seeds and skin obtained from some Serbian traditional grapes were outstanding with markedly high antioxidant potential.

2. Materials and Methods

2.1. Standards and Reagents

Ethanol, Trolox (97%), TPTZ (i.e., 2, 4, 6-tris(2-pyridyl)-s-triazine), DPPH (i.e., 2, 2′-diphenyl-1-picrylhydrazyl), ABTS (i.e., 2, 2′-azino-bis(3-ethylbenzthiazoline-6-sulphonic acid), choline chloride, glucose, citric acid. Analytical standards of quercetin 3-*O*-glucoside, (+)-catechin, (−)-epicatechin, protocatechuic acid, and gallic acid were obtained from Sigma (St. Luis, MO, USA). Hydrochloric acid, pyrogallol, *Folin-Ciocalteu* reagent (FC), potassium hydroxide, sodium carbonate, ferric chloride hexahydrate, and potassium peroxydisulfate were purchased from Merck (Darmstadt, Germany).

2.2. Preparation of NADES

Choline chloride (ChCl) was dried in the vacuum concentrator (Savant SPD131DDA SpeedVac Concentrator, Thermo scientific, USA) for 24 h at 60 °C. NADES solutions with 30% (*v/v*) of water were prepared by mixing water, ChCl and hydrogen bond donor (citric acid (Cit) or glucose (Glc)) at the respective molar ratio (ChCit = 2:1 and ChGlc = 1:1). The mixture was stirred in the sealed flask for 2 h at 50 °C until a homogeneous transparent uncolored liquid was formed.

2.3. Sample Preparation

Ten grape varieties (Table 1) were harvested in autumn 2017 at technological maturity. Immediately after collecting berries, seeds and skins were manually separated. Seeds were washed with cold, distilled water and air dried at room temperature until moisture content was below 10% (approximately three weeks). Crushed seeds were continuously extracted with chloroform for six hours at 70 °C in order to eliminate lipids from the samples. Skins were lyophilized and stored under nitrogen up to subsequent analysis. All samples were grinded into a powder form (BOSCH domestic coffee mill). Extraction was conducted by mixing samples with solvent of interest (70% acidified aqueous ethanol ChCit and ChGlc). The solid-liquid ratio was 1:10. Polyphenol-rich fraction was extracted from grape skins and defatted seeds using an ultrasound bath (FALC, Treviglio, Italy). The extraction parameters were: time 30 min, temperature 50 °C, and frequency 59 kHz [25,27]. Afterwards, mixtures were centrifuged (Janetzki T32 C, Wallhausen, Germany) for 15 min at 5000× *g* and supernatants were decanted to further analysis.

Table 1. Characteristics of grape varieties.

Grape Variety	Origin	Epoch of Maturation	Locality
'Chardonnay' (white)	i	II	3 Morave region- Trstenik
'Bagrina' (white)	a	III	Negotinska Krajina region- Negotin
'Župljanka' (white)	a	III	3 Morave region- Trstenik
'Gamay' (red)	i	II	3 Morave region- Trstenik
'Začinak' (red)	a	III	Negotinska Krajina region- Negotin
'Black Tamjanika' (red)	a	II	Negotinska Krajina region- Negotin
'Merlot' (red)	i	III	3 Morave region- Trstenik
'Prokupac' (red)	a	III/IV	3 Morave region- Trstenik
'Frankovka' (red)	i	II	3 Morave region- Trstenik
'Cabernet Sauvignon' (red)	i	III	3 Morave region- Trstenik

Abbreviations: i—international; a—autochthonous.

2.4. Determination of Total Polyphenol Content (TPC)

Total polyphenol content of extracts was determined using the rapid microtiter plate *Folin-Ciocalteu* method [28]. Briefly, 10 µL of diluted samples and serial standard solutions (gallic acid) were loaded on a 96-well microtiter plate (MTP). The repeated volumes of *Folin-Ciocalteu* reagent diluted 10 times (100 µL), and 1 M Na_2CO_3 (80 µL) were transferred to wells. After an hour of incubation at room temperature in the dark, the absorbance of blue coloration was measured at 630 nm against a blank sample on an MTP reader (BIOTEK, USA, ELx800 Absorbance Microplate Reader). The results were expressed as mg Gallic Acid Equivalents (GAE) per gram of dry weight (mg GAE g^{-1} DW). Calibration curve (y = 0.0028x + 0.0087), prepared for the working solutions of Gallic Acid in the concentration range of 10 to 80 mg L^{-1}, showed good linearity (r^2 = 0.9983). Limit of detection was 1.341 mg L^{-1}, and limit of quantification was 4.470 mg L^{-1}. Spectrophotometric analyses were done in triplicate.

2.5. Antioxidant Activity Evaluation

2.5.1. Ferric Ion Reducing Antioxidant Power (FRAP) Microassay

This test was performed according to Bolanos et al. with some modifications [29]. FRAP working solution was prepared by mixing 300 mM acetate buffer (pH = 3.6), 10 mM TPTZ solution (i.e., 2, 4, 6-tripyridyl-s-triazine in 40 mM HCl) and 20 mM $FeCl_3 \times 6H_2O$ at volume ratio 10:1:1. Diluted samples and Trolox solutions (20 µL) were added together with FRAP working solution (280 µL) in 96-well microplate in triplicate. Reaction mixtures were incubated at 37 °C for 30 min in dark conditions. The absorbance was measured at 630 nm. Antioxidant activity was calculated from calibration curve (y = 1.2368x + 0.0592) with the range 0.1–1 mmol Trolox L^{-1} and with linearity (r^2 = 0.9947). Limit of detection (LOD) was 6 µmol L^{-1}, while limit of quantification (LOQ) was 20 µmol L^{-1}. Results were expressed as mM Trolox Equivalents (TE) per gram of dry weight (mM TE g^{-1} DW).

2.5.2. Cupric Ion Reducing Antioxidant Capacity (CUPRAC) Microassay

CUPRAC assay was done as briefly described Zengin et al. [30]. Diluted extracts and Trolox solutions (67 µL) were pipetted in 96-well microplate in triplicate. Antioxidant capacity was evaluated after 30 min of reaction at room temperature between extracts and 61 µL 0.01 M $CuCl_2$, 61 µL 7.5×10^3 M neocuproine in ethanol and 61 µL ammonium acetate buffer (1 M, pH = 7). Absorbance readings were made against a reagent blank at 450 nm. Calibration curve (y = 1.4453x + 0.0467), calculated using a range 0.05–0.6 mmol Trolox L^{-1}, showed good linearity (r^2 = 0.9968). Limit of detection was 11 µmol L^{-1}, and limit of quantification was 37 µmol L^{-1}. Results were expressed as mM Trolox Equivalents (TE) per gram of dry weight (mM TE g^{-1} DW).

2.5.3. 2, 2'-Diphenyl-1-Picrylhydrazyl (DPPH) Radical Scavenging Microassay

Antioxidant activity was investigated using DPPH microassay as it was described by Melendez et al. with some modifications [31]. Diluted samples and standard solutions (7 µL) were allowed to react with 193 µL of the DPPH radical solution (1.86×10^{-4} mol L^{-1} DPPH in ethanol, prepared ex tempore) in 96-well microplate in triplicate. The mixtures were shaken and after incubation (1 h at room temperature); the absorbance was measured at 490 nm on an MTP reader. Trolox was used as a standard (range 0.2–1 mmol L^{-1}) for obtaining calibration curve (y = 62.691x − 0.112) with good linearity (r^2 = 0.993). LOD was 0.1 µmol L^{-1}, and LOQ was 0.4 µmol L^{-1}. Results were expressed as mM Trolox Equivalents (TE) per gram of dry weight (mM TE g^{-1} DW).

2.5.4. Trolox Equivalent Antioxidant Capacity (TEAC) Microassay

This method was based on ABTS (2, 2'-azino-bis(3-ethylbenzthiazoline-6-sulphonic acid)) radical decolorization adjusted to a microplate reader [32]. Stock solutions of ABTS (14 mM) and potassium peroxodisulfate (4.9 mM) in phosphate buffer (pH 7.4) were prepared and mixed in equal volumes.

The mixture was left at room temperature in the dark (1216 h) to allow free radical generation. On the day of analysis, the ABTS radical solution was diluted with phosphate buffer in order to achieve an absorbance of 0.7 ± 0.02 at 734 nm (approximately 1:80, *v/v*). The method was performed in the following way: diluted samples and Trolox solutions (20 µL) were added to 280 µL of the ABTS radical solution in wells of MTP in triplicate. Trolox solutions in the concentration range 0.02–0.15 mmol L^{-1} were used for calibration curve (y = 283.25x + 0.2318), with good linearity ($r^2 = 0.998$). LOD and LOQ were determined (0.1 and 0.5 µmol L^{-1}, respectively). After exactly 6 min, the absorbance readings were taken at 630 nm. Results were expressed as mM Trolox Equivalents (TE) per gram of dry weight (mM TE g^{-1} DW).

2.5.5. Antioxidant Composite Index (ACI)

ACI values were calculated in order to provide a comprehensive information about antioxidant activity of extracts. Combination of the results obtained from all performed tests (FRAP, CUPRAC, DPPH and ABTS) enabled the ACI calculation. An index value of 100 is assigned to the best score for each method and then an index score for all other samples was calculated following the equation:

$$\text{Antioxidant Index Score} = \frac{\text{sample score}}{\text{best score}} * 100 \tag{1}$$

The average of all four tests for each grape variety was then taken as antioxidant potency composite index [33].

2.6. HPLC Analysis

All samples were filtered through 0.22 µm polytetrafluoroethylene (PTFE) filter before the injection. HPLC analyses were performed on the Agilent 1200 Series HPLC system (Agilent, San Jose, CA, USA) equipped with a diode array detector (DAD) according to Panić et al., with small modifications [26]. Polyphenols were separated on a Phenomenex C18 column (Kinetex 150 mm × 4.6 mm, 2.6 µm, 100 Å) using the mobile phase, consisted of 2% HCOOH in H_2O (solvent A) and methanol (solvent B) at a flow rate of 0.8 mL/min. Gradient conditions were as follows: 3–8% B linear 0–13 min, 8% B isocratic 13–18 min, 8–10% B linear 18–20 min, 10% B isocratic 20–45 min, 10–100% B linear 45–50 min, 100% B isocratic 50–54 min, 100–3% B linear 54–55 min, 3% B isocratic 55–60 min. The wavelength used for the quantitative determination of catechins and phenolic acids was 280 nm, while flavonol derivatives were detected and quantified at 360 nm. Individual phenolic compounds were identified by comparing their spectral data and retention times with those of the authentic external standards. Calibration curve of external standard was used for their quantification. The calibration curves of standards were made by diluting the stock standards with methanol. Parameters of linear regression, LOD and LOQ for phenolic compounds by HPLC analysis are presented in Table 2.

Table 2. Parameters of linear regression, LOD and LOQ for phenolic compounds by HPLC analysis.

Compound	Concentration Range (mg L^{-1})	Regression Equation	r^2	LOD (mg L^{-1})	LOQ (mg L^{-1})
Gallic acid	0.22–60.60	34.008 x − 2.593	0.9996	0.198	0.696
Protocatechuic acid	0.24–490	34.025x + 15.663	0.9999	0.025	0.085
(+)-Catechin	0.43–56.67	7.832 x + 1.208	0.9999	0.191	0.637
(−)-Epicatechin	0.25–41.00	8.111 x − 1.309	0.9999	0.077	0.256
Quercetin 3-*O*-glucoside	0.19–22.40	35.138 x − 3.079	0.9998	0.069	0.231

Abbreviations: LOD: limit of detection; LOQ: limit of quantification.

2.7. Statistical Analysis

The statistical analysis was performed using the software SPSS (Version 20, Chicago, IL, USA). Homogeneity of variance was checked with Levene's test. Independent sample t- test was applied to evaluate differences between international and autochthonous grapes. Analysis of variance (ANOVA) was used to identify differences between varieties. Tukey's post-hoc test was used for multiple comparisons between groups. The relationship among all results was described by the Pearson's correlation coefficient. p value < 0.05 was considered statistically significant.

3. Results and Discussion

3.1. Total Polyphenol Content (TPC)

Total polyphenol content of grape seeds and skin was investigated in order to emphasize previous disclosures about the importance of profitable and sustainable utilization of wine by-products, considering appreciable amounts of biologically active compounds remained caught within the grape pomace [34,35]. The present study was conducted on seeds and skin from ten different grape varieties, among which, five of them represent Serbian autochthonous varieties ('Bagrina', 'Župljanka', 'Začinak', 'Black Tamjanika', and 'Prokupac'). With intention to compare conventional and natural deep eutectic solvents, investigation of total phenolic content was performed using three different solvents: acidified aqueous ethanol (AcEtOH), and green solvents ChCit, and ChGlc. Natural deep eutectic solvents are designer solvents, and there are more than 10^8 different possible combinations. Commonly, it is recommended to try several typical combinations, which considerably differ in physiochemical properties in order to select a suitable NADES. In our study, choline chloride-based NADES containing citric acid or glucose were chosen based on former reports and their physicochemical properties. For instance, ChCit is characterized with low pH value and with polarity similar to polarity of conventional solvents widely used for polyphenol extraction. Conversely, pH of ChGlc is almost neutral. High stability of phenolic compounds in sugar based NADES, as well as in ChCit, was previously published [10,14]. Furthermore, different water content in NADES should be optimized in order to reduce viscosity that can cause problem with mass transfer and pumping if the NADES are applied in industrial level [36]. In many cases, the addition of water between 20 30% (w/w) reduced viscosity of NADES and beneficially influences extraction yield of both polar and non-polar compounds. According to our experience and previous optimized works with polyphenol extraction, 30% of water was added to the NADES ChCit and ChGlc [25].

Choosing the right NADES for polyphenol extraction is a major part of the extraction process optimization, since it directly influences pH, polarity, and viscosity of a solvent, and as such, directly influences the extraction efficiency. However, NADES choice is followed by selection of the extraction method, which also significantly contributes to the extraction efficiency. In our case, ultrasound assisted extraction (UAE) was chosen since extraction yield could be improved with cavitation phenomena resulting cell leakage and better mass transfer. Additionally, process parameters to achieve the maximum extraction efficiency within NADES should be adjusted. This means optimization of a temperature, extraction time, and power of ultrasound. According to our experience, and previously published and optimized ultrasound assisted extractions of polyphenols from the grape within NADES, extraction was carried out 30 min on 50 °C [25].

TPC values of grape seed extracts obtained from acidified aqueous ethanol ranged from 70.86 to 146.69 mg GAE per gram of dry weight (Figure 1). The highest total phenolic content was extracted from 'Gamay' and 'Prokupac' seeds, while lowest TPC appeared in seeds obtained from 'Merlot' and 'Cabernet Sauvignon' varieties. Our results were similar to those reported for the same varieties [18,22] and slightly lower than seed data obtained for Mediterranean grape pomace by Ky et al. [37]. despite of the similarity of the varieties included in our and in the study conducted by Pantelic et al., minor observed differences are presumably present due to seasonal variations of weather, agronomical practices, and vineyard management [22]. Nevertheless, mentioned study also reported the highest total polyphenol content in seeds obtained from Serbian autochthonous variety 'Prokupac'. Regarding NADES, TPC values of grape seed extracts ranged from 92.36 to 182.60 and 56.17 to 156.17 mg GAE per gram of dry weight for ChCit and ChGlc extracts, respectively (Figure 1). ChCit had higher extraction efficiency than ChGlc in almost all varieties (with the exception of seeds of two red grape varieties: 'Merlot' and 'Prokupac'). Interestingly, seeds from three Serbian autochthonous varieties, 'Prokupac', 'Black Tamjanika' and 'Župljanka', stood out with pronounced high TPC content in both NADES.

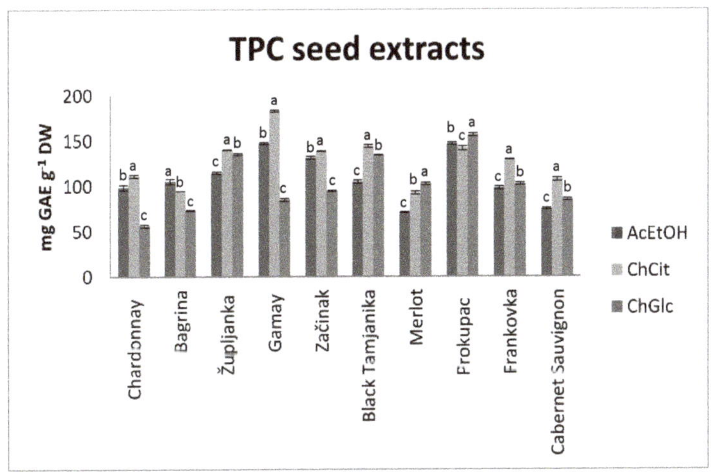

Figure 1. Total polyphenol content of seed extracts obtained from ten grape varieties. Data are expressed as mg GAE g^{-1} DW for each variety, means ± SD, $n = 3$. Different lower case letters (a–c) indicate significant differences among solvents ($p \leq 0.05$) based on post hoc Tukey's test. Abbreviations: TPC—total polyphenol content; GAE—gallic acid equivalents; DW—dry weight; SD—standard deviation; AcEtOH—acidified aqueous ethanol; ChCit—choline chloride: citric acid; ChGlc—choline chloride: glucose.

By analyzing skin extracts (Figure 2) obtained using acidified aqueous ethanol, 'Merlot', 'Cabernet Sauvignon', and 'Začinak' were outstanding with the highest TPC (16.21, 16.14, and 15.00 mg GAE g^{-1} of DW, respectively). Concerning only white varieties, 'Župljanka' (9.41 mg GAE g^{-1} of DW) had the highest total phenolic content. The results of TPC in skin extracts revealed in the present study were in good agreement with literature [18,23]. When it comes to NADES skin extracts (Figure 2), ChCit outperformed ChGlc, with some exceptions. ChCit and acidified aqueous ethanol showed similar potential considering solvent extractability; therefore, the same varieties ('Cabernet Sauvignon', 'Merlot', and 'Začinak') had the highest total phenolic content in both extracts. ChGlc distinguished those, but also 'Frankovka', as grape varieties, which skin is abundant in phenolic compounds. The present study confirmed the observations of Cvjetko-Bubalo et al., who screened five different DES and proposed acid based NADES as the most promising for extraction of grape skin polyphenols [25].

In general, TPC values varied significantly among varieties (Figures 1 and 2). In addition, it is clearly observed that total phenolics of grape skin extracts were several times lower than those determined for grape seed extracts. These findings were in accordance with previous reported data for the *Vitis vinifera* species [18]. In addition, TPC values determined for red skin were slightly higher in comparison with white skin extracts.

Extracting solvent affected total polyphenol content of both seeds and skins (Figures 1 and 2). It is known that physicochemical properties play an important role in solvents selectivity towards some particular compounds. More polar solvent is expected to have a higher yield of polar molecules in comparison with non-polar ones. Thus, differences in extraction efficiency between ChCit and ChGlc could be present due to distinction in their polarity (acid based NADES are more polar). Solvent acidity is another important feature found to affect extraction selectivity. For example, anthocyanins are flavonoid phenolic compounds whose chemical form and stability depend on pH value. Namely, they are prevalent in the flavylium cation form, which is stable at low pH. With the increased pH value, they undergo structural transformation into forms prone to further degradation due to their instability [25,38]. Therefore, extraction of these polyphenols is enhanced along with the growing acidity of a solvent. This could be a possible explanation for lower TPC values found in most ChGlc extracts, considering almost neutral pH of this solvent (pH values of AcEtOH and ChCit are both less than 7). ChCit showed equal or even higher extraction efficiency than referent solvent, acidified aqueous ethanol, observing both, seeds and skin. Extraordinary extraction capacity of ChCit extracts confirmed previous findings about high extraction ability of acid based NADES for some phenolic compounds [25,38–40]. Moreover, acidified aqueous ethanol and ChCit extracts were significantly correlated (Pearson's coefficient of 0.792 and 0.950; $p < 0.01$, for seeds and skin, respectively); due to similar acidity and polarity of the solvents.

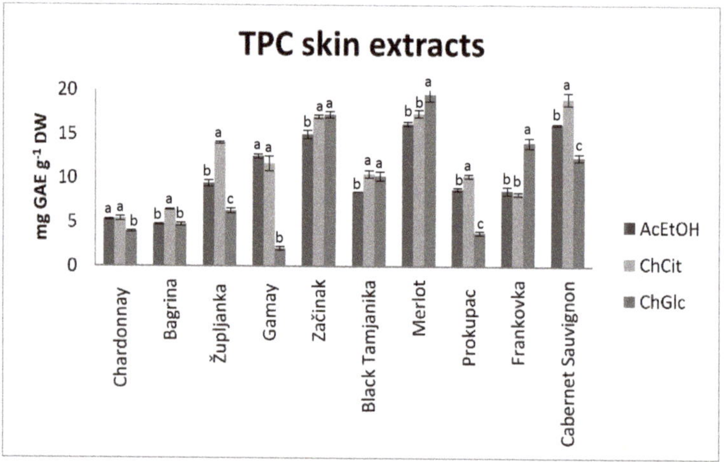

Figure 2. Total polyphenol content of skin extracts obtained from ten grape varieties. Data are expressed as mg GAE g^{-1} DW for each variety, means ± SD, $n = 3$. Different lower case letters (a–c) indicate significant differences among solvents ($p \leq 0.05$) based on post hoc Tukey's test. Abbreviations: TPC—total polyphenol content; GAE—gallic acid equivalents; DW—dry weight; SD—standard deviation; AcEtOH—acidified aqueous ethanol; ChCit—choline chloride: citric acid; ChGlc—choline chloride: glucose.

3.2. Antioxidant Activity

Dietary antioxidants, widely spread in fruits and vegetables, have received considerable scientific attention and investigation of their bioactivity became the main goal of many conducted studies. Regardless of the large number of validated antioxidant in in vitro tests, there is still a problem concerning the differences in chemical principles and mechanisms of the most common methods. In order to provide comprehensive information about total antioxidant capacity, and to distinguish the dominant mechanism of antioxidant activity, it was suggested to perform several tests pointing different aspects of antioxidant properties [41]. Therefore, four antioxidant methods were conducted—FRAP, CUPRAC, DPPH, and ABTS (Table S1: Antioxidant activity of grape seed extracts, Table S2: Antioxidant activity of grape skin extracts).

FRAP is a typical electron transfer method based on the capability of antioxidants to reduce Fe^{3+} tripyridyl triazine complex (colorless) to Fe^{2+} tripyridyl triazine (blue color) in acidic medium. CUPRAC assay measures the ability of test compounds to reduce Cu^{2+} to Cu^+ in aqueous-ethanolic medium (pH 7.0) in the presence of neocuproine. DPPH assay, the oldest indirect method for evaluating antioxidant capacity, measures the reduction of stable organic DPPH radical in presence of antioxidants to the yellow colored 2, 2-diphenyl-1-picrylhydrazyn. Another commonly used radical scavenging test is ABTS/TEAC assay, based on the ability of test compounds to reduce ABTS radical [41].

Taking into account many limitations and lacks of each test applied for evaluating antioxidant activity, results obtained from them were used for calculating unique antioxidant composite index (ACI). ACI is a simple, widely used mathematical calculation that enables deeper insight into antioxidant activity, since it combines several tests based on different chemical mechanisms.

ACI values calculated for grape seed extracts are presented in Figure 3. Regarding conventional solvents, the highest antioxidant activity was determined for seeds obtained from 'Gamay' variety. Extraction with ChCit also revealed 'Gamay' as the most potent, followed by Serbian autochthonous varieties, such as 'Black Tamjanika' and 'Prokupac'. In the case of ChGlc, the highest ACI value was calculated for seed extracts of 'Prokupac' variety. When it comes to grape skin (Figure 4), ACI calculation revealed apparently red skin prominence in antioxidant actions. 'Merlot', 'Začinak', and 'Cabernet Sauvignon' were characterized with the highest antioxidant capacity. Observed differences among varieties, regarding both seeds and skin, represent a confirmation of many research studies conducted so far [20,42].

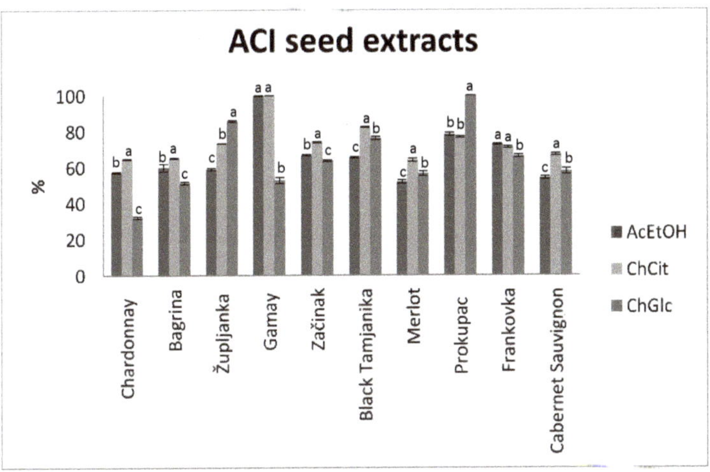

Figure 3. Antioxidant composite index- grape seed extracts.Data are expressed as % for each variety, means ± SD, n = 3. Different lower case letters (a–c) indicate significant differences among solvents ($p \leq 0.05$) based on post hoc Tukey's test. Abbreviations: ACI—Antioxidant Composite Index; SD—standard deviation; AcEtOH—acidified aqueous ethanol; ChCit—choline chloride: citric acid; ChGlc—choline chloride: glucose.

ACI values for both, seeds and skin (Figures 3 and 4) were significantly correlated with TPC of corresponding extracts ($0.798 \leq r \leq 0.967; p < 0.01$). Therefore, such strong relationship confirmed the previous disclosures about the significant contribution of polyphenol compounds to the antioxidant activity. Nevertheless, prominent antioxidant activity of ChCit extracts was revealed. Higher antioxidant activity of these extracts in comparison with ChGlc and acidified aqueous ethanol extracts could be explained by the antioxidant activity of ChCit itself, or by antioxidant activity of NADES forming compound, citric acid. This naturally occurring organic acid possess antioxidant and anti-inflammatory properties. Furthermore, antitumor activity of this compound has been reported [43]. Citric acid is widely used food additive, approved and closely regulated as acidulant, pH regulator, flavoring agent, preservative and antioxidant synergist in soft drinks, baked nutrients, jam, candy, jelly sweet, marmalade, and tinned vegetable [44].

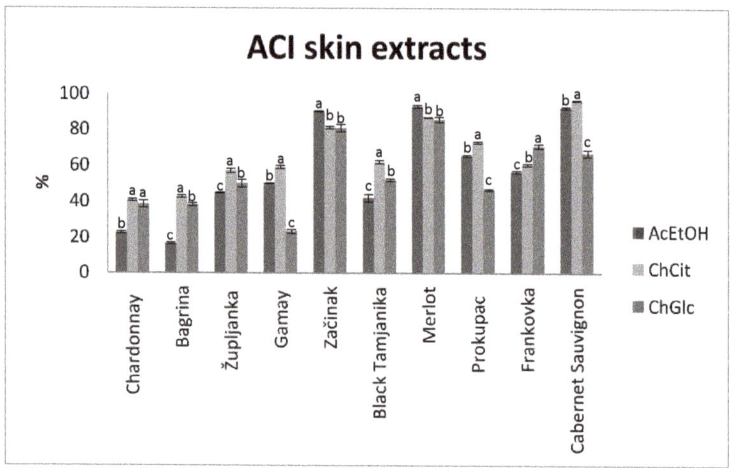

Figure 4. Antioxidant composite index- grape skin extracts. Data are expressed as % for each variety, means ± SD, $n = 3$. Different lower case letters (a–c) indicate significant differences among solvents ($p \leq 0.05$) based on post hoc Tukey's test. Abbreviations: ACI—Antioxidant Composite Index; SD—standard deviation; AcEtOH—acidified aqueous ethanol; ChCit—choline chloride: citric acid; ChGlc—choline chloride: glucose.

3.3. Phenolic Composition

Structural and physicochemical features of a solvent have a strong influence on selectivity towards biologically active compounds. In general, considering solvent extractability, the highest content of polyphenols from both, seeds and skin, was obtained with ChCit. Thus, further HPLC analysis was done to precisely characterize the phenolic profile of such ChCit extracts. Moreover, acidified aqueous ethanol extracts were also screened in order to compare NADES versus conventional solvent in terms of extraction efficiency for some of the main grape polyphenols (gallic acid and protocatechuic acid within the phenolic acids group, (+)-catechin and (−)-epicatechin as flavan-3-ols representatives, quercetin 3-*O*-glucoside like flavonol main member).

The sum of identified polyphenols by HPLC varied noticeably among varieties, for both seed and skin extracts. Polyphenol compositions evaluated by HPLC-DAD analysis of grape seed extracts are presented in Table 3. Flavan-3-ols were predominantly present, which is in accordance with previously published data [22,45]. (+)-Catechin was the most abundant followed by (−)-epicatechin. Gallic acid and protocatechuic acid were present in smaller quantities. When it comes to grape skin (Table 4), distribution of phenolic compounds was significantly different amid varieties, which is in accordance with literature data [46]. Nevertheless, grape skin peaks could be assigned to the four different compounds: (+)-catechin, (−)-epicatechin, protocatechuic acid, and quercetin 3-*O*-glucoside.

Regardless of the similar polarity and acidity of extraction solvents, differences between acidified aqueous ethanol and ChCit selectivity were observed. Conventional solvent, almost without the exception, had a higher affinity than NADES for (+)-catechin, (−)-epicatechin, and gallic acid in seeds. The highest concentrations of (+)-catechin were determined in the seeds of 'Župljanka' (15.587 and 10.197 mg g^{-1} DW, for acidified aqueous ethanol and ChCit, respectively), followed by 'Začinak' and 'Prokupac'. 'Prokupac' was the variety with the highest content of (−)-epicatechin, regardless the solvent. The abundance of dominant flavan-3-ols was in agreement with Ky et al. [34] and Bakkalbaşı et al. [47]. Gallic acid was predominantly present hydroxybenzoic acids. Interestingly, there was no difference between seeds obtained from white and red varieties regarding gallic acid content, although such observation was previously reported [21,48]. Among all varieties, red grape 'Prokupac' showed significantly higher amounts of gallic acid (2.450 and 1.850 mg g^{-1} DW, for acidified aqueous ethanol and ChCit seed extracts, respectively).

Table 3. Phenolic composition of grape seed extracts.

Grape Variety	Solvent	Gallic Acid (mg g^{-1} DW)	Protocatechuic Acid (mg g^{-1} DW)	(+)-Catechin (mg g^{-1} DW)	(−)-Epicatechin (mg g^{-1} DW)	Total (mg g^{-1} DW)
'Chardonnay'	AcEtOH	0.947 [e]		3.661 [d]	1.737 [d]	6.345 [d]
	ChCit	0.885 [e]	0.482 [b]	3.347 [d]	1.466 [d]	6.181 [d]
'Bagrina'	AcEtOH	1.204 [d]	0.158 [e]	4.355 [d]	3.605 [c]	9.322 [c]
	ChCit	1.142 [d]	0.249 [d]	3.045 [d]	1.626 [d]	6.062 [d]
'Župljanka'	AcEtOH	1.546 [c]	0.569 [a]	15.587 [a]	4.391 [b]	22.093 [a]
	ChCit	1.219 [d]	0.605 [a]	10.197 [b]	2.808 [d]	14.829 [b]
'Gamay'	AcEtOH	0.689 [f]	0.101 [e]	2.911 [d]	2.371 [d]	6.073 [d]
	ChCit	1.096 [d]	0.291 [c]	3.998 [d]	1.785 [d]	7.169 [d]
'Začinak'	AcEtOH	1.230 [d]	0.115 [e]	6.884 [c]	2.189 [d]	10.419 [c]
	ChCit	1.167 [d]	0.214 [e]	4.883 [d]	0.948 [e]	7.213 [d]
'Black Tamjanika'	AcEtOH	1.691 [c]	0.228 [e]	4.937 [d]	3.550 [c]	10.406 [c]
	ChCit	1.519 [c]	0.444 [b]	4.248 [d]	2.292 [d]	8.504 [d]
'Merlot'	AcEtOH	0.992 [e]	0.140 [e]	3.799 [d]	2.356 [d]	7.287 [d]
	ChClt	0.940 [e]	0.276 [d]	3.020 [d]	1.275 [e]	5.510 [d]
'Prokupac'	AcEtOH	2.450 [a]	0.219 [e]	6.709 [c]	6.269 [a]	15.647 [b]
	ChCit	1.850 [b]	0.383 [c]	5.338 [d]	2.999 [d]	10.570 [c]
'Frankovka'	AcEtOH	1.908 [b]	0.522 [b]	5.224 [d]	5.140 [b]	12.794 [c]
	ChCit	1.315 [d]	0.508 [b]	3.281 [d]	2.705 [d]	7.810 [d]
'Cabernet Sauvignon'	AcEtOH	0.786 [f]	0.135 [e]	3.960 [d]	1.797 [e]	6.679 [d]
	ChCit	0.745 [f]	0.262 [d]	3.177 [d]	0.970 [e]	5.154 [d]

Values represent mean of three replicates. Standard deviation was < 5%. Different lower case letters (a–f) indicate significant differences among solvents and varieties ($p \leq 0.05$) based on post hoc Tukey's test. Abbreviations: AcEtOH—acidified aqueous ethanol, ChCit—choline chloride: citric acid; DW—dry weight.

Table 4. Phenolic composition of grape skin extracts.

Grape Variety	Solvent	Protocatechuic Acid (mg g^{-1} DW)	(+)-Catechin (mg g^{-1} DW)	(−)-Epicatechin (mg g^{-1} DW)	Quercetin 3-O-Glucoside (mg g^{-1} DW)	Total (mg g^{-1} DW)
'Chardonnay'	AcEtOH		0.240 [b]			0.240 [e]
	ChCit	0.562 [b]	0.245 [b]			0.807 [e]
'Bagrina'	AcEtOH	0.139 [c]	0.227 [b]			0.366 [e]
	ChCit	0.293 [c]	0.294 [b]			0.587 [e]
'Župljanka'	AcEtOH		0.277 [b]			0.277 [e]
	ChCit	0.147 [c]	0.275 [b]	0.035 [d]		0.457 [e]
'Gamay'	AcEtOH			0.429 [c]	0.115 [e]	0.544 [e]
	ChCit	0.330 [c]	0.146 [d]	0.689 [c]		1.165 [d]
'Začinak'	AcEtOH		0.095 [e]	2.479 [b]	0.312 [d]	2.886 [c]
	ChCit	1.663 [a]	0.191 [c]	2.768 [b]	0.274 [d]	4.896 [b]
'Black Tamjanika'	AcEtOH				0.348 [d]	0.348 [e]
	ChCit	0.275 [c]	0.061 [e]	0.032 [d]		0.369 [e]
'Merlot'	AcEtOH		0.060 [e]	2.066 [b]		2.127 [c]
	ChCit	0.085 [c]	0.265 [b]	4.654 [a]	0.334 [d]	5.338 [b]
'Prokupac'	AcEtOH		0.061 [e]	0.068 [d]	0.187 [e]	0.316 [e]
	ChCit	0.166 [c]	0.252 [b]	0.222 [c]	0.130 [e]	0.770 [e]
'Frankovka'	AcEtOH		0.085 [e]	0.106 [c]	0.420 [c]	0.611 [e]
	ChCit	0.065 [c]	0.307 [a]	0.329 [c]		0.701 [e]
'Cabernet Sauvignon'	AcEtOH		0.017 [e]	2.612 [b]	0.569 [b]	3.198 [c]
	ChCit	0.098 [c]	0.180 [c]	5.219 [a]	0.739 [a]	6.237 [a]

Values represent mean of three replicates. Standard deviation was < 5%. Different lower case letters (a–e) indicate significant differences among solvents and varieties ($p \leq 0.05$) based on post hoc Tukey's test. Abbreviations: AcEtOH—acidified aqueous ethanol, ChCit—choline chloride: citric acid; DW—dry weight.

Interestingly, ChCit skin extracts contained higher amounts of (+)-catechin and (−)-epicatechin in comparison with acidified aqueous ethanol extracts. (−)-Epicatechin was dominantly present in red grape skin and the highest concentrations of this compound were determined in 'Cabernet Sauvignon' (2.612 and 5.219 mg g^{-1} DW, for acidified aqueous ethanol and ChCit extracts, respectively). Among white varieties, isomer (+)-catechin was the most abundant.

ChCit also showed better extraction efficiency for protocatechuic acid irrespective of investigated matrix. Concerning seeds, 'Župljanka' was characterized with the highest concentration of protocatechuic acid (0.569 and 0.605 mg g^{-1} DW, for acidified aqueous ethanol and ChCit, respectively). When it comes to skin, acidified aqueous ethanol demonstrated extremely low extraction efficiency towards protocatechuic acid. Concentrations of protocatechuic acid in ChCit skin extracts varied from 0.065 to 1.663 mg g^{-1} DW.

The concentration of quercetin 3-*O*-glucoside, found in skin extracts obtained from red grape varieties, tended to be higher in acidified aqueous ethanol compared to NADES, with some exceptions. The highest concentration of this compound was observed in skin of 'Cabernet Sauvignon' (0.569 and 0.739 mg g^{-1} DW, for acidified aqueous ethanol and ChCit, respectively).

In general, 'Župljanka' seeds distinguished with the highest content of total phenolics evaluated applying HPLC, regardless the solvent. Such domination of 'Župljanka' among the varieties with the highest TPC content determined by *Folin-Ciocalteu* method, such as 'Gamay', 'Prokupac', and 'Black Tamjanika' is due to remarkably higher (+)-catechin concentrations in 'Župljanka' seeds. Thus, it is expected that mentioned varieties have higher concentrations of some unquantified polyphenol compounds in seeds. When it comes to grape skin, highly pigmented red varieties, such as 'Cabernet Sauvignon', 'Začinak', and 'Merlot' could be emphasized as cultivars extremely rich in phenolic compounds.

Finally, chromatographic analysis of obtained extracts revealed some interesting observations. Namely, acidified aqueous ethanol was more efficient in extracting phenolics from grape seeds, while ChCit have proved more selective towards target polyphenol compounds in grape skin. Considering the domination of ChCit in TPC determination, such lack of consistency in results obtained from HPLC analyses could be explained by the presence of some phenolic compounds that we were not able to quantify. For example, flavanols derivatives and polymers such as: (−)-epicatechin gallate, epigallocatechin gallate, and proanthocyanidins have been identified in both seeds and skin extracts, although seeds contained significantly higher amounts. Perhaps the quantification of these compounds would shed a new light on a phenolic composition of grape seeds, since it was previously reported that acid based NADES possess an excellent extraction performance for catechins [49].

The food industries add value both by reducing waste disposal and by transforming by-products into new value food items. Based on obtained results in this study, we would like to suggest utilization of grape pomace as a good source of biologically active compounds. Recovery of polyphenols from grape waste appears to be a new strategy in commercialized applications especially using ecological and sustainable extraction.

4. Conclusions

Considering a diverse array of biologically active compounds present in grape, valorization of grape pomace by extracting potent antioxidants represents a great challenge. This study investigated antioxidant potential and polyphenol composition of seeds and skin obtained from some widely grown grapes with an emphasis on Serbian old autochthonous varieties. International grapes did not surpass autochthonous ones concerning parameters of interest. Moreover, some Serbian varieties distinguished as potent sources of polyphenols. For example, high polyphenol concentrations were determined in 'Prokupac' and 'Black Tamjanika' seeds. Furthermore, variety 'Župljanka' was outstanding with significantly higher amounts of catechins in seeds. When it comes to skin, noteworthy TPC values were determined for 'Začinak' variety. Therefore, broaden exploitation of seeds and skin from Serbian traditional varieties is strongly recommended.

Additionally, because of a growing awareness about environmental pollution, there is a need to minimize and even eliminate the use of hazardous chemicals in extraction processes. Within this research, it was shown that conventional solvents could be replaced with natural deep eutectic solvents, yielding the same or better polyphenol content. NADES were reported as safe for human consumption, and they could be used in the industry without difficult and expensive downstream purification steps.

Since ChCit exhibited high extraction efficiency towards polyphenols from grape seeds and skin, the present work strongly emphasizes its utilization for green and sustainable extraction of biologically active compounds from wine industry pomace.

Supplementary Materials: The following are available online at http://www.mdpi.com/2076-3417/10/14/4830/s1, Table S1: Antioxidant activity of grape seed extracts, Table S2: Antioxidant activity of grape skin extracts, Figure S1: HPLC chromatogram of polyphenols in a sample of grape seeds, Figure S2: HPLC chromatogram of polyphenols in a sample of grape skin.

Author Contributions: Conceptualization, I.R.R. and S.Š.; methodology, V.T. and M.P.; investigation, N.D. and M.P.; writing-original draft preparation, N.D.; writing—review and editing, N.D., V.T.; visualization, I.R.R. and S.Š.; supervision, I.R.R. and S.Š. All authors have read and agreed to the published version of the manuscript.

Funding: This research was funded by Serbian Ministry of Education, Science, and Technological Development (451-03-68/2020-14/200161).

Acknowledgments: We gratefully thank winery Milosavljevic Bucje, winery Matalj Negotin, and Agricultural High School Rajko Bosnic Negotin for their kind donation of grape samples.

Conflicts of Interest: The authors declare no conflict of interest. The funders had no role in the design of the study; in the collection, analyses, or interpretation of data; in the writing of the manuscript, or in the decision to publish the results.

References

1. Bail, S.; Stuebiger, G.; Krist, S.; Unterweger, H.; Buchbauer, G. Characterisation of various grape seed oils by volatile compounds, triacylglycerol composition, total phenols and antioxidant capacity. *Food Chem.* **2008**, *108*, 1122–1132. [CrossRef] [PubMed]

2. Adams, D.O. Phenolics and ripening in grape berries. *Am. J. Enol. Vitic.* **2006**, *57*, 249–256.

3. Nassiri-Asl, M.; Hosseinzadeh, H. Review of the pharmacological effects of Vitis vinifera (Grape) and its bioactive constituents: An update. *Phytother. Res.* **2016**, *30*, 1392–1403. [CrossRef] [PubMed]

4. Rasines-Perea, Z.; Teissedre, P.L. Grape polyphenols' effects in human cardiovascular diseases and diabetes. *Molecules* **2017**, *22*, 68. [CrossRef]

5. Calapai, G.; Bonina, F.; Bonina, A.; Rizza, L.; Mannucci, C.; Arcoraci, V.; Laganà, G.; Alibrandi, A.; Pollicino, C.; Inferrera, S.; et al. A Randomized, double-blinded, clinical trial on effects of a Vitis vinifera extract on cognitive function in healthy older adults. *Front. Pharmacol.* **2017**, *8*, 776. [CrossRef] [PubMed]

6. FAOSTAT. Food and Agriculture Organization of the United Nations. Available online: http://www.fao.org/faostat/en/#home (accessed on 24 February 2020).

7. Rudra, S.G.; Nishad, J.; Jakhar, N.; Kaur, C. Food industry waste: Mine of nutraceuticals. *Int. J. Sci. Environ. Technol.* **2015**, *4*, 205–229.

8. Maier, T.; Schieber, A.; Kammerer, D.R.; Carle, R. Residues of grape (Vitis vinifera L.) seed oil production as a valuable source of phenolic antioxidants. *Food Chem.* **2009**, *112*, 551–559. [CrossRef]

9. Beres, C.; Costa, G.N.; Cabezudo, I.; da Silva-James, N.K.; Teles, A.S.; Cruz, A.P.; Mellinger-Silva, C.; Tonon, R.V.; Cabral, L.M.; Freitas, S.P. Towards integral utilization of grape pomace from winemaking process: A review. *Waste Manag.* **2017**, *68*, 581–594. [CrossRef]

10. Panić, M.; Gunjević, V.; Cravotto, G.; Redovniković, I.R. Enabling technologies for the extraction of grape-pomace anthocyanins using natural deep eutectic solvents in up-to-half-litre batches extraction of grape-pomace anthocyanins using NADES. *Food Chem.* **2019**, *300*, 125185. [CrossRef]

11. Skarpalezos, D.; Detsi, A. Deep Eutectic Solvents as Extraction Media for Valuable Flavonoids from Natural Sources. *Appl. Sci.* **2019**, *9*, 4169. [CrossRef]

12. Radošević, K.; Ćurko, N.; Srček, V.G.; Bubalo, M.C.; Tomašević, M.; Ganić, K.K.; Redovniković, I.R. Natural deep eutectic solvents as beneficial extractants for enhancement of plant extracts bioactivity. *LWT Food Sci. Technol.* **2016**, *73*, 45–51. [CrossRef]

13. Aroso, I.M.; Paiva, A.; Reis, R.L.; Duarte, A.R.C. Natural deep eutectic solvents from choline chloride and betaine- Physicochemical properties. *J. Mol. Liq.* **2017**, *241*, 654–661. [CrossRef]

14. Dai, Y.; Verpoorte, R.; Choi, Y.H. Natural deep eutectic solvents providing enhanced stability of natural colorants from safflower (Carthamus tinctorius). *Food Chem.* **2014**, *159*, 116–121. [CrossRef] [PubMed]

15. Gorke, J.T.; Srienc, F.; Kazlauskas, R.J. Hydrolase-catalyzed biotransformations in deep eutectic solvents. *Chem. Commun.* **2008**, *10*, 1235–1237. [CrossRef] [PubMed]

16. Kaar, J.L.; Jesionowski, A.M.; Berberich, J.A.; Moulton, R.; Russell, A.J. Impact of ionic liquid physical properties on lipase activity and stability. *J. Am. Chem. Soc.* **2003**, *125*, 4125–4131. [CrossRef] [PubMed]

17. Kragl, U.; Eckstein, M.; Kaftzik, N. Enzyme catalysis in ionic liquids. *Curr. Opin. Biotechnol.* **2002**, *13*, 565–571. [CrossRef]

18. Xu, C.; Zhang, Y.; Cao, L.; Lu, J. Phenolic compounds and antioxidant properties of different grape cultivars grown in China. *Food Chem.* **2010**, *119*, 1557–1565. [CrossRef]

19. Obreque-Slier, E.; Peña-Neira, A.; Lopez-Solis, R.; Zamora-Marín, F.; Ricardo-da Silva, J.M.; Laureano, O. Comparative study of the phenolic composition of seeds and skins from Carménère and 'Cabernet Sauvignon' grape varieties (Vitis vinifera L.) during ripening. *J. Agric. Food Chem.* **2010**, *58*, 3591–3599. [CrossRef]

20. Katalinić, V.; Možina, S.S.; Skroza, D.; Generalić, I.; Abramovič, H.; Miloš, M.; Ljubenkov, I.; Piskernik, S.; Pezo, I.; Terpinc, P.; et al. Polyphenolic profile, antioxidant properties and antimicrobial activity of grape skin extracts of 14 Vitis vinifera varieties grown in Dalmatia (Croatia). *Food Chem.* **2010**, *119*, 715–723. [CrossRef]

21. Godevac, D.; Tešević, V.; Velickovic, M.; Vujisić, L.V.; Vajs, V.; Milosavljević, S. Polyphenolic compounds in seeds from some grape cultivars grown in Serbia. *J. Serb. Chem. Soc.* **2010**, *75*, 1641–1652. [CrossRef]

22. Pantelić, M.M.; Zagorac, D.Č.D.; Davidović, S.M.; Todić, S.R.; Bešlić, Z.S.; Gašić, U.M.; Tešić, Ž.L.; Natić, M.M. Identification and quantification of phenolic compounds in berry skin, pulp, and seeds in 13 grapevine varieties grown in Serbia. *Food Chem.* **2016**, *211*, 243–252. [CrossRef]

23. Šuković, D.; Knežević, B.; Gašić, U.; Sredojević, M.; Ćirić, I.; Todić, S.; Mutić, J.; Tešić, Ž. Phenolic Profiles of Leaves, Grapes and Wine of Grapevine Variety Vranac (Vitis vinifera L.) from Montenegro. *Foods* **2020**, *9*, 138. [CrossRef] [PubMed]

24. Zdunić, G.; Godevac, D.; Šavikin, K.; Krivokuća, D.; Mihailović, M.; Pržić, Z.; Marković, N. Grape seed polyphenols and fatty acids of autochthonous 'Prokupac' vine variety from Serbia. *Chem. Biodivers.* **2019**, *16*, 1900053. [CrossRef]

25. Bubalo, M.C.; Ćurko, N.; Tomašević, M.; Ganić, K.K.; Redovniković, I.R. Green extraction of grape skin phenolics by using deep eutectic solvents. *Food Chem.* **2016**, *200*, 159–166. [CrossRef] [PubMed]

26. Panić, M.; Stojković, M.R.; Kraljić, K.; Škevin, D.; Redovniković, I.R.; Srček, V.G.; Radošević, K. Ready-to-use green polyphenolic extracts from food by-products. *Food Chem.* **2019**, *283*, 628–636. [CrossRef] [PubMed]

27. Todorovic, V.; Milenkovic, M.; Vidovic, B.; Todorovic, Z.; Sobajic, S. Correlation between antimicrobial, antioxidant activity, and polyphenols of alkalized/nonalkalized cocoa powders. *J. Food Sci.* **2017**, *82*, 1020–1027. [CrossRef] [PubMed]

28. Attard, E. A rapid microtitre plate Folin-Ciocalteu method for the assessment of polyphenols. *Open Life Sci.* **2013**, *8*, 48–53. [CrossRef]

29. Bolanos de la Torre, A.A.S.; Henderson, T.; Nigam, P.S.; Owusu-Apenten, R.K. A universally calibrated microplate ferric reducing antioxidant power (FRAP) assay for foods and applications to Manuka honey. *Food Chem.* **2015**, *174*, 119–123. [CrossRef]

30. Zengin, G.; Sarikurkcu, C.; Aktumsek, A.; Ceylan, R. Sideritis galatica Bornm.: A source of multifunctional agents for the management of oxidative damage, Alzheimer's's and diabetes mellitus. *J. Funct. Foods* **2014**, *11*, 538–547. [CrossRef]

31. Meléndez, N.P.; Nevárez-Moorillón, V.; Rodríguez-Herrera, R.; Espinoza, J.C.; Aguilar, C.N. A microassay for quantification of 2, 2-diphenyl-1-picrylhydracyl (DPPH) free radical scavenging. *Afr. J. Biochem. Res.* **2014**, *8*, 14–18. [CrossRef]

32. Pastoriza, S.; Delgado-Andrade, C.; Haro, A.; Rufián-Henares, J.A. A physiologic approach to test the global antioxidant response of foods. The GAR method. *Food Chem.* **2011**, *129*, 1926–1932. [CrossRef]

33. Seeram, N.P.; Aviram, M.; Zhang, Y.; Henning, S.M.; Feng, L.; Dreher, M.; Heber, D. Comparison of antioxidant potency of commonly consumed polyphenol-rich beverages in the United States. *J. Agric. Food Chem.* **2008**, *56*, 1415–1422. [CrossRef] [PubMed]

34. Ky, I.; Lorrain, B.; Kolbas, N.; Crozier, A.; Teissedre, P.L. Wine by-products: Phenolic characterization and antioxidant activity evaluation of grapes and grape pomaces from six different French grape varieties. *Molecules* **2014**, *19*, 482. [CrossRef]

35. Rockenbach, I.I.; Gonzaga, L.V.; Rizelio, V.M.; Gonçalves, A.E.D.S.S.; Genovese, M.I.; Fett, R. Phenolic compounds and antioxidant activity of seed and skin extracts of red grape (Vitis vinifera and Vitis labrusca) pomace from Brazilian winemaking. *Food Res. Int.* **2011**, *44*, 897–901. [CrossRef]
36. Hayyan, A.; Mjalli, F.S.; AlNashef, I.M.; Al-Wahaibi, Y.M.; Al-Wahaibi, T.; Hashim, M.A. Glucose-based deep eutectic solvents: Physical properties. *J. Mol. Liq.* **2013**, *178*, 137–141. [CrossRef]
37. Ky, I.; Teissedre, P.L. Characterisation of Mediterranean grape pomace seed and skin extracts: Polyphenolic content and antioxidant activity. *Molecules* **2015**, *20*, 2190. [CrossRef]
38. Bosiljkov, T.; Dujmić, F.; Bubalo, M.C.; Hribar, J.; Vidrih, R.; Brnčić, M.; Zlatic, E.; Redovniković, I.R.; Jokić, S. Natural deep eutectic solvents and ultrasound-assisted extraction: Green approaches for extraction of wine lees anthocyanins. *Food Bioprod. Process.* **2017**, *102*, 195–203. [CrossRef]
39. Cao, J.; Chen, L.; Li, M.; Cao, F.; Zhao, L.; Su, E. Efficient extraction of proanthocyanidin from Ginkgo biloba leaves employing rationally designed deep eutectic solvent-water mixture and evaluation of the antioxidant activity. *J. Pharm. Biomed. Anal.* **2018**, *158*, 317–326. [CrossRef]
40. Ruesgas-Ramón, M.; Figueroa-Espinoza, M.C.; Durand, E. Application of deep eutectic solvents (DES) for phenolic compounds extraction: Overview, challenges, and opportunities. *J. Agric. Food Chem.* **2017**, *65*, 3591–3601. [CrossRef] [PubMed]
41. Gulcin, İ. Antioxidants and antioxidant methods: An updated overview. *Arch. Toxicol.* **2020**, *94*, 651–715. [CrossRef]
42. Baiano, A.; Terracone, C. Varietal differences among the phenolic profiles and antioxidant activities of seven table grape cultivars grown in the south of Italy based on chemometrics. *J. Agric. Food Chem.* **2011**, *59*, 9815–9826. [CrossRef] [PubMed]
43. Chen, X.; Lv, Q.; Liu, Y.; Deng, W. Effect of food additive citric acid on the growth of human esophageal carcinoma cell line EC109. *Cell J.* **2017**, *18*, 493–502. [CrossRef] [PubMed]
44. Yılmaz, S.; Ünal, F.; Yüzbaşıoğlu, D.; Aksoy, H. Clastogenic effects of food additive citric acid in human peripheral lymphocytes. *Cytotechnology* **2008**, *56*, 137–144. [CrossRef] [PubMed]
45. Di Lecce, G.; Arranz, S.; Jauregui, O.; Tresserra-Rimbau, A.; Quifer-Rada, P.; Lamuela-Raventos, R.M. Phenolic profiling of the skin, pulp and seeds of Albariño grapes using hybrid quadrupole time-of-flight and triple-quadrupole mass spectrometry. *Food Chem.* **2014**, *145*, 874–882. [CrossRef] [PubMed]
46. González-Manzano, S.; Rivas-Gonzalo, J.C.; Santos-Buelga, C. Extraction of flavan-3-ols from grape seed and skin into wine using simulated maceration. *Anal. Chim. Acta* **2004**, *513*, 283–289. [CrossRef]
47. Bakkalbaşı, E.; Yemiş, O.; Aslanova, D.; Artık, N. Major flavan-3-ol composition and antioxidant activity of seeds from different grape cultivars grown in Turkey. *Eur. Food Res. Technol.* **2005**, *221*, 792–797. [CrossRef]
48. Montealegre, R.R.; Peces, R.R.; Vozmediano, J.C.; Gascueña, J.M.; Romero, E.G. Phenolic compounds in skins and seeds of ten grape Vitis vinifera varieties grown in a warm climate. *J. Food Compost. Anal.* **2006**, *19*, 687–693. [CrossRef]
49. Li, J.; Han, Z.; Yu, B. Efficient extraction of major catechins in Camellia sinensis leaves using green choline chloride-based deep eutectic solvents. *RSC Adv.* **2015**, *5*, 93937–93944. [CrossRef]

 *applied sciences*

Article

Development of a Low-Temperature and High-Performance Green Extraction Process for the Recovery of Polyphenolic Phytochemicals from Waste Potato Peels Using Hydroxypropyl β-Cyclodextrin

Achillia Lakka, Stavros Lalas and Dimitris P. Makris *

Green Processes & Biorefinery Group, School of Agricultural Sciences, University of Thessaly, N. Temponera Str., 43100 Karditsa, Greece; ACHLAKKA@uth.gr (A.L.); slalas@uth.gr (S.L.)
* Correspondence: dimitrismakris@uth.gr; Tel.: +30-24410-64792

Received: 12 May 2020; Accepted: 21 May 2020; Published: 23 May 2020

Abstract: Potato peels (PP) are a major agri-food side-stream originating from potato processing, but to date, their green valorization as a bioresource of antioxidant polyphenols is limited to extraction processes involving mainly water/ethanol-based solvents, whereas other eco-friendly methodologies are scarce. This study aimed at developing a simple, straight-forward and green extraction methodology to effectively recover PP polyphenols, using hydroxypropyl β-cyclodextrin (HP-β-CD). After an initial assay to identify the optimal HP-β-CD concentration that would provide increased extraction yield, optimization based on response surface methodology enabled maximization of the extraction performance, providing a total polyphenol yield of 17.27 ± 0.93 mg chlorogenic acid equivalent g^{-1} dry mass, at 30 °C. Testing of temperatures higher than 30 °C and up to 80 °C did not favor higher yields. The extracts obtained with HP-β-CD were slightly richer in polyphenols than extracts prepared with conventional solvents, such as aqueous ethanol and methanol, displaying similar antioxidant characteristics. The major polyphenols that could be identified in the extracts were neochlorogenic, chlorogenic, caffeic and ferulic acids. The outcome of this study demonstrated that HP-β-CD may be used as a highly effective green means of recovering PP polyphenols, at near-ambient temperature.

Keywords: antioxidants; extraction; hydroxypropyl β-cyclodextrin; polyphenols; potato peels

---

1. Introduction

Industrial activity within the agri-food sector is a major source of side-streams and wastes, which pose serious and imminent environmental risks. However, the dire necessity for proper waste management and handling does not stem merely from avoiding environmental aggravation, but also from the recognition that waste agri-food biomass is a primal bioresource of raw materials destined for the production of value-added substances [1]. Various plant tissues regularly rejected during processing of agricultural products to foods, such as stems, peels, seeds, leaves etc., constitute highly important, abundant and low-cost pools of precious phytochemicals, whose use may have wide applicability in food, pharmaceutical and cosmetic formulations [2–4].

Potato (*Solanum tuberosum*) is an indispensable element in the nutritional habits of numerous countries around the globe, and there have been estimations that world production of potato amounted 368 million tons in 2013. This makes potatoes the fourth largest food crop, after rice, wheat, and maize [5]. There is an enormous amount of by-products generated through potato tuber processing, since potatoes are regularly used to produce a spectrum of popular food commodities, such as chips, canned potatoes, mashed potatoes, fries etc. Potato processing residues are composed almost exclusively of peels, which are regarded as a source of polyphenolic substances with significant nutritional value [6,7].

The effective recovery of bioactive polyphenols from various waste plant materials has traditionally been based on the use of common volatile solvents of appropriate polarity, such as methanol, ethanol and acetone. However, such approaches are now considered as options lower in the hierarchy of biorefinery strategies, as the principles of Green Chemistry explicitly dictate the use of alternative green solvents, possessing negligible vapor pressure, recyclability and no toxicity [8]. In this framework, the seeking and testing of novel extraction media is of undisputed significance. Cyclodextrins (CDs) are cycling supramolecular structures, containing at least 6 D-(+)-glucopyranoside units, attached to each other with α-1,4-glycosidic linkages [9]. The main feature characterizing CDs is their relatively hydrophobic cavity, as opposed to their external surface, which is hydrophilic. This unique structural attribute permits CDs to form inclusion complexes with compounds possessing a variety of structures, which are stabilized through non-covalent forces, such as hydrophobic interactions, van der Waals forces and hydrogen bonds [10].

Cyclodextrins have been extensively used in recent years as means of assisting extraction of polyphenolic compounds, and several studies have appraised factors critical to an extraction process, such as the CD type, CD concentration, time, extraction technique and temperature [11]. The advantages ascribed to the use of CDs for extracting polyphenols pertain mainly to the improvement of extraction efficiency and extraction time shortening, display of improved antioxidant activity by the extracts produced, and better extract stabilization and solute bioavailability. In this frame, the current study was performed to examine polyphenol extraction from PP with the aid of HP-β-CD, and identify key process variables, whose optimization could contribute to attaining maximized extraction yield.

2. Materials and Methods

2.1. Chemicals and Reagents

Neochlorogenic acid (≥98%) was from Merck (Darmstadt, Germany). Chlorogenic acid (95%) was from Fluorochem (Hadfield, UK). Ferulic acid, caffeic acid, Folin–Ciocalteu reagent, 2,4,6-tris(2-pyridyl)-*s*-triazine (TPTZ), hydroxypropyl β-cyclodextrin and 2,2-diphenylpicrylhydrazyl (DPPH) were from Sigma-Aldrich (Darmstadt, Germany). Sodium hydroxide and iron chloride hexahydrate was from Merck (Darmstadt, Germany). Sodium carbonate anhydrous (99%), aluminum chloride anhydrous (98%) and ethanol (99.8%) were from Penta (Praha, Czechia). All solvents used for chromatography were HPLC grade.

2.2. Plant Material

Waste potato peels (PP) with a moisture level of 82%, derived from processing of brown-skin potatoes (*Solanum tuberosum* L. var Spunta), were collected from a catering facility (Karditsa, Greece) and transferred to the laboratory within 15 min after peeling. PP were then immediately placed on discs to form layers of approximate thickness of 0.5 cm, and dried in a laboratory oven (Binder BD56, Bohemia, NY, USA) for 480 min at 75 °C, to give a moisture content of approximately 4%. The dried PP were pulverized in a ball mill and sieved to provide a powder with a mean particle diameter of 0.435 mm. This feed was used for all extractions.

2.3. Batch Stirred-Tank Solid–Liquid Extraction Process

To test the effect of hydroxypropyl β-cyclodextrin (HP-β-CD) concentration (C_{CD}) prior to optimization, accurately weighted PP amount of 0.667 g was transferred into a 50 mL round-bottom flask and mixed with 20 mL deionized water containing varying concentrations of HP-β-CD (0–8 mM). Extractions were performed at room temperature (29 ± 2 °C) on a magnetic stirrer (Witeg, Wertheim, Germany) at 500 rpm for 150 min. After extraction, samples were centrifuged at 10,000× g and stored at −40 °C. For the extractions carried out for the response surface optimization, the amount of PP and the stirring speed varied according to predetermined values, dictated by experimental design. The pH of the extraction medium was adjusted at various levels with 2 M NaOH, after incorporating citric acid at a final concentration of 1 g L^{-1}. Extractions with 60% (*v/v*) EtOH and 60% (*v/v*) MeOH were used as the control.

2.4. Experimental Design and Process Optimization

Process optimization was accomplished by deploying response surface methodology through a Box–Behnken design with three central points. The objective was to evaluate the influence of selected extraction variables on the total polyphenol yield (Y_{TP}), which was chosen as the response. The variables examined were the pH of the extraction medium, the liquid-to-solid ratio ($R_{L/S}$) and the stirring speed (S_S). These variables were assigned as X_1, X_2 and X_3, respectively (Table 1), and codified to levels, as previously reported in detail [12]. The assessment of the fitting of the model to the experimental data was based on ANOVA and lack-of-fit analysis. The equation describing the mathematical model was reported by excluding non-significant dependent terms.

Table 1. Actual and coded levels of the process variables considered for response surface optimization.

Independent Variables	Code Units	Coded Variable Level		
		−1	0	1
pH	X_1	2	3.5	5
$R_{L/S}$ (mL g^{-1})	X_2	20	50	80
S_S (rpm)	X_3	200	500	800

2.5. Total Polyphenol Determination

Total polyphenol determination was performed using the Folin–Ciocalteu methodology, as described elsewhere [12]. In an Eppendorf tube of 1.5 mL, a 0.1 mL sample followed by 0.1 mL Folin–Ciocalteu reagent were introduced and allowed to stand for 2 min. Then, 0.8 mL of sodium carbonate (5%) was added and the mixture was maintained for 20 min, at 40 °C, in a water bath (Heidolph HB digital, Schwabach, Germany). Results were given as mg chlorogenic acid equivalents per g dry mass (dm).

2.6. Estimation of the Antiradical Activity (A_{AR})

For this assay, DPPH was used as the radical probe and a stoichiometric determination was employed [12]. A volume of 0.975 mL DPPH (100 μM in methanol) was combined with 0.025 mL of sample and incubated at room temperature. The absorbance at 515 nm was recorded at $t = 0$ min ($A_{515(i)}$) and at $t = 30$ min ($A_{515(f)}$). The A_{AR} was computed as follows:

$$A_{AR} = \frac{\Delta A}{\varepsilon \times l \times C} \times Y_{TP} \tag{1}$$

ΔA corresponds to $A_{515(i)} - A_{515(f)}$, ε (DPPH) to $11,126 \times 10^6$ μM^{-1} cm^{-1}, C to $C_{TP} \times 0.025 \times$ sample dilution (in this case 1/50), Y_{TP} to the total polyphenol extraction yield (mg g^{-1}) and l to the path length (1 cm). A_{AR} was given as μmol DPPH g^{-1} dm.

2.7. Estimation of the Ferric-Reducing Power (P_R)

A published methodology was implemented to assay P_R [12]. In short, 0.05 mL of FeCl$_3$ (4 mM in 0.05 M HCl) and an equal volume of extract were mixed and incubated for 30 min at 37 °C, in a water bath. Following this, 0.9 mL of TPTZ solution (1 mM in 0.05 M HCl) was added, and the absorbance was recorded at 620 nm, after 10 min. P_R was estimated as μM ascorbic acid equivalents (AAE) per g dry mass.

2.8. High-Performance Liquid Chromatography-Diode Array (HPLC-DAD)

The equipment employed was a Shimadzu CBM-20A liquid chromatograph (Shimadzu Europa GmbH, Duisburg, Germany), coupled to a Shimadzu SPD-M20A detector, and interfaced by Shimadzu LC solution software. For the chromatographic analyses, a Phenomenex Luna C18(2) column (100 Å, 5 µm, 4.6 × 250 mm) (Phenomenex, Inc., Torrance, CA, USA) was used, at 40 °C. Separation was performed with eluents (A) 0.5% aqueous formic acid and (B) 0.5% formic acid in MeCN/water (6:4), at 1 mL min^{-1} flow rate, using the following elution program: 100% A to 60% A in 40 min, 60% A to 50% A in 10 min, 50% A to 30% A in 10 min; then, isocratic elution for further 10 min. The injection volume was 20 µL. Quantification was performed at 325 nm, with calibration curves (0–50 µg mL^{-1}), constructed with neochlorogenic acid ($R^2 = 0.998$), chlorogenic acid ($R^2 = 0.999$), caffeic acid ($R^2 = 0.998$) and ferulic acid ($R^2 = 0.998$).

2.9. Statistical Processing

Extractions were repeated at least twice and analyses in triplicate. The reported values are means ± standard deviation (sd). SigmaPlot™ 12.5 (Systat Software Inc., San Jose, CA, USA) was used to establish linear correlations, at least at a 95% significance level. JMP™ Pro 13 (SAS, Cary, NC, USA) was used to carry out distribution analysis, as well as to set up the response surface experimental design and perform the response surface-associated statistics (ANOVA, lack-of-fit).

3. Results and Discussion

3.1. Effect of HP-β-CD Concentration (C_{CD})

The initial step in the process development was to identify the most suitable C_{CD} which would provide the highest total polyphenol extraction yield (Y_{TP}). Thus, C_{CD} was tested over a range of 0 (deionized water) to 8 mM (Figure 1). Compared to deionized water (0 mM), a solution with C_{CD} of 0.1 mM provided by 14.5% higher Y_{TP}. This fact indicated that the presence of HP-β-CD fostered polyphenol extraction to a significant extent. By raising C_{CD} levels from 0.1 to 0.25 mM, the Y_{TP} increased from 9.76 ± 0.20 to 9.84 ± 0.87 mg CGAE g^{-1} dm. This difference fell within the limits of statistical error and it was non-significant ($p > 0.05$). However, switching C_{CD} from 0.25 to 0.5 mM, an increase in Y_{TP} by 19.1% was recorded, whereas a further C_{CD} increase up to 8 mM afforded higher Y_{TP} by only 2%. Therefore, a C_{CD} of 0.5 mM was chosen as the most appropriate to carry out subsequent process optimization.

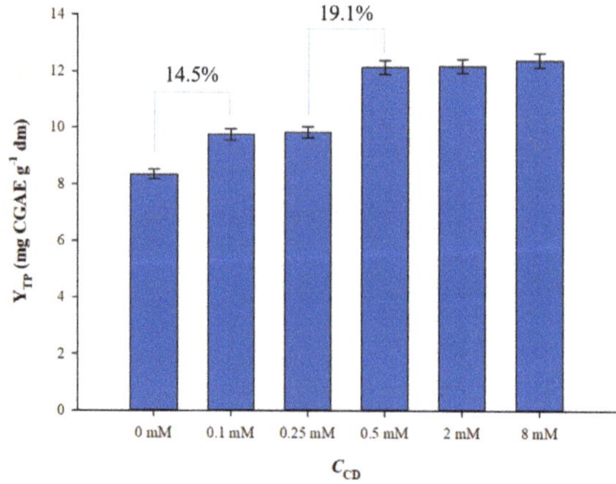

Figure 1. Effect of HP-β-CD concentration (C_{CD}) on total polyphenol yield (Y_{TP}). Extractions were carried out at ambient temperature (29 ± 2 °C), at a liquid-to-solid ratio of 30 mL g^{-1}.

3.2. Optimization-Effect of Process Variables

The optimization procedure was designed to evaluate the influence exerted by three critical extraction variables, pH, $R_{L/S}$ and S_S, on the response (Y_{TP}), and to reveal possible synergistic effects between them. The assessment of the fitted model and the suitability of response surface were accomplished by performing ANOVA and a lack-of-fit test (Figure 2), on the basis of the proximity of measured and predicted values (Table 2). Taking into consideration only the significant terms (Parameter Estimates, Figure 2), the second-degree mathematical model derived was as follows:

$$Y_{TP} \text{ (mg CGAE g}^{-1} \text{ dm)} = 11.35 + 2.62X_2 + 0.32X_3 + 1.30X_1{}^2 + 1.36X_3{}^2 \tag{2}$$

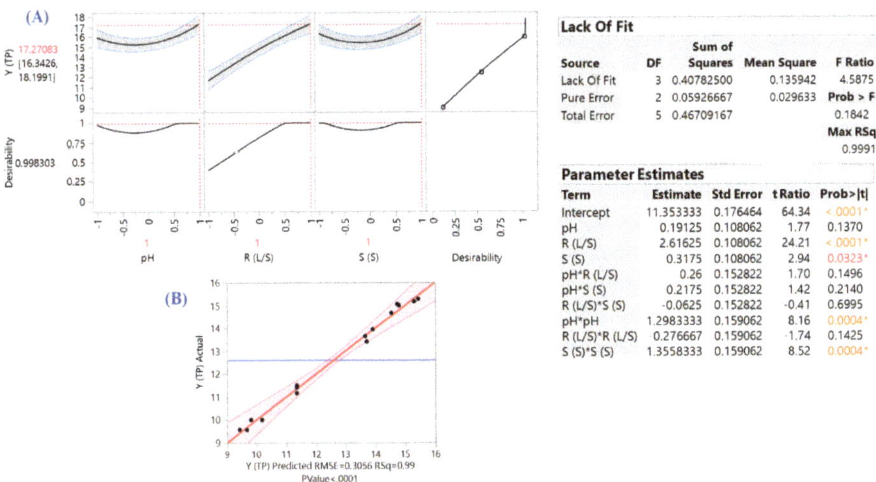

Lack Of Fit

Source	DF	Sum of Squares	Mean Square	F Ratio
Lack Of Fit	3	0.40782500	0.135942	4.5875
Pure Error	2	0.05926667	0.029633	Prob > F
Total Error	5	0.46709167		0.1842
				Max RSq
				0.9991

Parameter Estimates

Term	Estimate	Std Error	t Ratio	Prob>|t|
Intercept	11.353333	0.176464	64.34	<.0001*
pH	0.19125	0.108062	1.77	0.1370
R (L/S)	2.61625	0.108062	24.21	<.0001*
S (S)	0.3175	0.108062	2.94	0.0323*
pH*R (L/S)	0.26	0.152822	1.70	0.1496
pH*S (S)	0.2175	0.152822	1.42	0.2140
R (L/S)*S (S)	-0.0625	0.152822	-0.41	0.6995
pH*pH	1.2983333	0.159062	8.16	0.0004*
R (L/S)*R (L/S)	0.276667	0.159062	-1.74	0.1425
S (S)*S (S)	1.3558333	0.159062	8.52	0.0004*

Figure 2. Desirability function (**A**), actual-by-predicted plot (**B**) and statistical information (lack-of-fit and parameter estimates inset tables) generated by performing response surface methodology, concerning the effect of process variables on the response. Asterisks (*) on the "Parameter Estimates" inset table denote statistically significant terms ($p < 0.05$).

Table 2. Analytical display of individual design points and measured and predicted values of the response.

Design Point	Independent Variables				Response (Y_{TP}, mg CGAE g^{-1} dw)
	X_1	X_2	X_3	Measured	Predicted
1	−1	−1	0	10.00	9.83
2	−1	1	0	14.66	14.54
3	1	−1	0	9.57	9.69
4	1	1	0	15.27	15.44
5	0	−1	−1	9.58	9.44
6	0	−1	1	10.00	10.20
7	0	1	−1	14.99	14.79
8	0	1	1	15.16	15.30
9	−1	0	−1	13.4	13.72
10	1	0	−1	13.64	13.66
11	−1	0	1	13.94	13.92
12	1	0	1	15.05	14.73
13	0	0	0	11.41	11.35
14	0	0	0	11.16	11.35
15	0	0	0	11.49	11.35

The square correlation coefficient (R^2) provides an indication of the total variability around the mean, as determined by the model [13]. Given that R^2 was 0.99 ($p < 0.0001$) and the lack-of-fit was significant at least at a 95% significance level, it could be supported that the mathematical equation (Equation (2)) describing the model showed excellent adjustment to the experimental data. The 3D diagrams representing the model (Figure 3) visualize the effect of the process variables on the response (Y_{TP}).

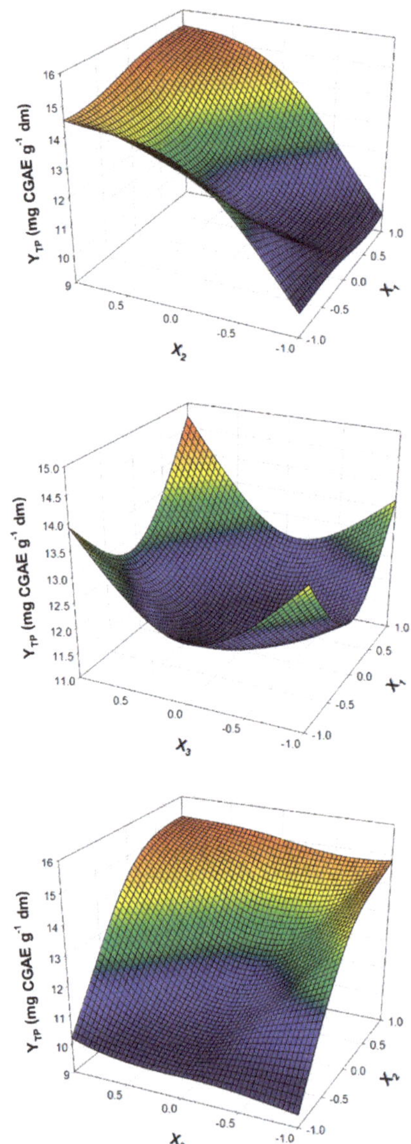

Figure 3. Three-dimensional diagrams portraying the effect of process variables on the response.

Using the desirability function (Figure 2), the maximum Y_{TP} was estimated to be 17.27 ± 0.93 mg CGAE g^{-1} dm, at the following optimal settings: pH = 5.0, $R_{L/S}$ = 80 mL g^{-1} and S_S = 800 rpm. To confirm the validity of this prediction, the optimal settings were employed to carry out three individual extractions and the Y_{TP} found was 16.86 ± 1.75 mg CGAE g^{-1} dm. This finding demonstrated the reliability of the predictive model. The Y_{TP} level determined in this study was of a similar magnitude to 20 mg GAE g^{-1} dm attained by employing subcritical water extraction of PP [14], but exceeded significantly 7.67 mg GAE g^{-1} dm achieved with 80% methanol and ultrasound-assisted extraction [15], 9.11 mg CAE g^{-1} dm achieved with 59% (*v/v*) ethanol and ultrasound-assisted extraction [16], and 11.0 mg GAE g^{-1} dm achieved with 60% ethanol and microwave-assisted extraction [17]. Lower yields of 6.4 mg g^{-1} dm obtained with 55% ethanol and ultrasound-assisted extraction [18], and 4.39 mg GAE g^{-1} dm obtained with pressurized water/ethanol mixtures [19] have also been reported. Although the total polyphenol content of PP may exhibit variations attributed to varietal, cultivation and climatic factors, in a survey on 60 different potato cultivars, no value higher than 12.59 mg GAE g^{-1} dm was found for the total polyphenol content [20]. This fact strongly evidenced that the level of Y_{TP} achieved in this study was indicative of a high-performance extraction system.

The direct effect of pH (X_1) was non-significant, but the influence exerted by pH was manifested by its quadratic term X_1^2. The optimum pH 5 determined by the model was the upper limit of the experimental design, suggesting that extraction was not favored in a more acidic environment. The role of pH has been implicated in polyphenol cyclodextrin interactions because it affects the dissociation of either carboxyl functions or phenolic hydroxyls. Experiments with caffeic acid (CA) and HP-β-CD indicated that constants of complex formation were sensitive to pH and decreased with pH increases [21]. This behavior was ascribed to the hydrophobicity of CA, which depends on its ionization state. Since the main inclusion interactions are hydrophobic, developed between the HP-β-CD cavity and the guest molecule, acidic pH would contribute to suppressing CA ionization, maintaining it in a non-ionized and thus, more hydrophobic form, hence, the greater stability constant. Results from rosmarinic acid/β-CD interactions were in concurrence [22]. However, the carboxyl group of the quinic moiety on CGA, which is the principal polyphenol in PP [5], has a pK_a of 3.33, and at pH 5, this group would be deprotonated [23]. Therefore, it would be expected that polyphenol extraction from PP with HP-β-CD would be favored at lower pH. This finding might indicate that other interactions between CGA and HP-β-CD and/or interactions with other PP constituents affected the overall extraction yield.

The effect of $R_{L/S}$ and S_S was significant (Figure 2, inset table), as demonstrated for several extraction processes [24–27]. Both variables have been closely associated with diffusion phenomena that govern solid–liquid extraction, and their proper regulation might be critical in attaining high extraction yield. Regarding $R_{L/S}$, it defines the concentration gradient that is the driving force of diffusion, which in turn, favors mass transfer [28–30]. Thus, suitable setting of $R_{L/S}$ would entail negligible resistance to mass transfer and effective entrainment of solute (polyphenols) into the liquid medium [30]. In this study, the optimum $R_{L/S}$ estimated was 80 mL g^{-1}, which matches the optimum 82–84 mL g^{-1} found for polyphenol extraction from PP and eggplant peels using water/ethanol mixtures [16,31].

Likewise, careful regulation of S_S results in effective dispersion of the solute that diffuses from the internal part of the solid particles, into the liquid volume. This would minimize external resistance and increase mass transfer [32]. To achieve higher diffusivity and increased mass transfer, S_S must be properly set, since it has been shown that diffusivity is highly correlated with S_S [33]. The optimum S_S of 800 rpm determined in this study was equal to those estimated for polyphenol extraction with deep eutectic solvent from saffron processing wastes [26] and *Moringa oleifera* leaves [27], and comparable to 900 rpm found for polyphenol extraction from onion solid wastes [34]. However, optimum S_S of as low as 200 rpm has also been reported [35].

3.3. Effect of Extraction Temperature

The extraction carried out at ambient *T* (approximately 30 °C) was demonstrated to be the most efficacious ($p < 0.05$), giving a Y_{TP} of 16.86 ± 1.75 mg CGAE g^{-1} dm, whereas extractions accomplished between 40 and 80 °C were of lower performance (Figure 4). This fact strongly emphasized that

$T > 30\ °C$ may not facilitate PP polyphenol extraction in the presence of HP-β-CD. Although a similar T of 35 °C has been proposed as the optimum for ultrasound-assisted PP polyphenol extraction [18], this finding contrasted with earlier investigations, which illustrated that the most favorable T for polyphenol extraction from PP may lie from 80 [16,17,19] and 90 [36] to as high as 190 °C [14]. However, intermediate levels of 50 °C have also been suggested [37].

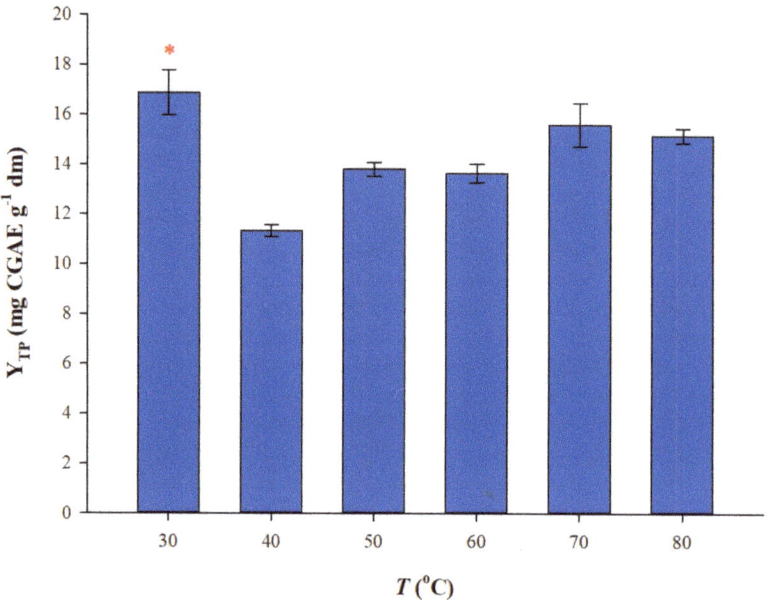

Figure 4. The effect of temperature on total polyphenol yield (Y_{TP}), under optimized extraction conditions. Asterisk (*) denotes statistically different value ($p < 0.05$). All other values showed no statistical difference.

The aqueous solubility of sparingly soluble polyphenols, such as quercetin, has been explicitly shown to increase with increases in T, in the presence of various cyclodextrins [38]. On this ground, the results illustrated in Figure 4 might appear paradoxical.

However, in another study on polyphenol/cyclodextrin complexation, it was evidenced that the binding constants of polyphenol complexes with α-, β- and γ-cyclodextrin, decreased with a rise in T [39]. In support of this finding were results from another investigation on interactions of the stilbene (E)-piceatannol with HP-β-CD, where a decrease in the complexation constant was observed as a response to raising T [40]. The authors ascribed this phenomenon to weakening of hydrogen bonds between the host and the guest molecule, due to heating. This hypothesis might justify the reason for reduced Y_{TP} observed at $T > 30\ °C$. Moreover, at T around 50–60 °C, there may be observed decomposition of polyphenol-CD inclusion complexes. Thus, selection of appropriate extraction T is of utmost importance [11].

3.4. Antioxidant Activity and Polyphenolic Composition

To better portray the effect of HP-β-CD on polyphenol extraction from PP, the extracts obtained under optimized conditions, at 30 °C, were tested for antioxidant activity. The results were compared with extracts prepared using 60% aqueous ethanol, a commonly used green solvent, and 60% aqueous methanol, a common conventional solvent (Table 3).

Table 3. Comparison of the characteristics of the extract obtained with HP-β-CD with those of common conventional solvents. Values reported are means ± standard deviation.

Extraction	Y_{TP} (mg CGAE g^{-1} dm)	A_{AR} (μmol DPPH g^{-1} dm)	P_R (μmol AAE g^{-1} dm)
HP-β-CD	16.86 ± 1.75	16.37 ± 0.49	8.83 ± 0.09
60% EtOH	13.67 ± 0.27	25.26 ± 0.76	9.47 ± 0.09
60% MeOH	13.27 ± 1.01	23.04 ± 0.69	11.49 ± 0.11

Y_{TP}, yield in total polyphenols; A_{AR}, antiradical activity; P_R, ferric-reducing power.

Extraction with HP-β-CD afforded almost by 19% higher Y_{TP} compared to 60% EtOH and by 21% compared to 60% MeOH, while differences in both A_{AR} and P_R were marginal and non-significant ($p > 0.05$). This outcome clearly suggested that extraction of PP with HP-β-CD was virtually as effective as with common organic solvents. The HP-β-CD extract was then analyzed with HPLC to identify the major polyphenolic constituents. Of the trace recorded at 325 nm (Figure 5), four phytochemicals could be tentatively identified, namely neochlorogenic acid, chlorogenic acid, caffeic acid and ferulic acid.

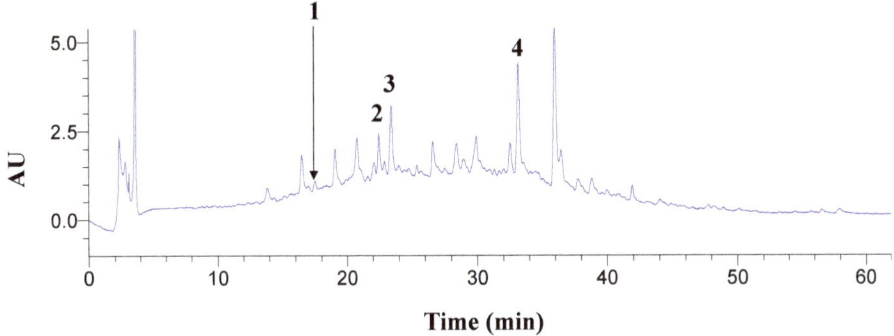

Figure 5. Chromatogram of PP extract obtained with HP-β-CD, under optimized conditions, at 30 °C. The trace was recorded at 325 nm. Peak assignment: 1, neochlorogenic acid; 2, chlorogenic acid; 3, caffeic acid; 4, ferulic acid.

The results from the quantitative analysis are given in Table 4. The most abundant polyphenol was ferulic acid, followed by caffeic and chlorogenic acid, whereas neochlorogenic acid occurred at much lower levels. Unlike previous reports, chlorogenic and caffeic acids were not the predominant constituents. In fact, chlorogenic acid content (83.67 μg g^{-1} dm) was significantly low compared to 4100 μg g^{-1} dm reported by other studies [18]. Yet, other authors determined levels ranging from 2.16 to 267.4 μg g^{-1} dm [15] and from 299.51 to 904.21 μg g^{-1} dm [41].

Table 4. Contents of the major polyphenols detected in PP extracts.

Peak #	Polyphenol	Content (μg g^{-1} dm) ± sd
1	Neochlorogenic acid	28.95 ± 1.60
2	Chlorogenic acid	83.67 ± 0.54
3	Caffeic acid	88.98 ± 0.28
4	Ferulic acid	108.73 ± 3.52
	Sum	310.34

Regarding caffeic acid, the content determined was comparable to 68.19–129.05 μg g^{-1} dm found in earlier investigations [15]. However, a yield of 592 μg g^{-1} dm has been achieved for PP extraction with pressurized liquids [19], but yields varying from 160 to as high as 1220 μg g^{-1} dm have been attained with ethanol/water mixtures and ultrasound-assisted extraction [18]. Other authors reported values of 250 to 324 $\mu g\ g^{-1}$ dm [41]. On the other hand, ferulic acid was shown to occur at levels of 1 to 28 $\mu g\ g^{-1}$ dm [42].

4. Conclusions

The examination presented herein illustrated the effectiveness of PP polyphenol extraction, at near-ambient temperatures, using HP-β-CD. Such an extraction process for PP polyphenols has not been reported heretofore. The results obtained suggested that regulation of pH at 5 may enhance polyphenol extractability, but the role of S_S and $R_{L/S}$ was also significant. Temperatures higher than 30 °C were not favorable to attaining higher extraction yields. Under optimized conditions, the total polyphenol yield and the antioxidant properties of the extracts produced were comparable to those obtained with aqueous methanol and ethanol. Some of the major polyphenols identified in the extracts were neochlorogenic, chlorogenic, caffeic and ferulic acids. The methodology proposed might be less cost-effective compared to methods employing cheaper conventional solvents. Another limitation that may arise by the use of HP-β-CD could be its incompatibility with certain types of foods and/or cosmetics (e.g., those with high-fat content). On the other hand, it is a completely volatile solvent-free process, and thus, the use of toxic and environmentally aggravating materials is fully precluded. Furthermore, HP-β-CD is a food-grade substance and the direct application of extracts produced with HP-β-CD, in foods and/or cosmetics, could be particularly advantageous, without the need for additional energy-demanding and time-consuming downstream steps for extract purification. It would be proposed that future studies should focus on identifying whether extracts generated with the method proposed could be effective food antioxidants/antimicrobials and/or cosmetic constituents. Such approaches would certainly have an obvious industrial prospect for large-scale valorization of PP as abundant and low-cost bioresources of polyphenolic phytochemicals.

Author Contributions: Conceptualization, S.L. and D.P.M.; formal analysis, A.L., S.L., and D.P.M.; investigation, A.L., SL., and D.P.M.; writing—original draft preparation, S.L. and D.P.M.; writing—review and editing, S.L. and D.P.M.; supervision, D.P.M. All authors have read and agreed to the published version of the manuscript.

Funding: This research received no external funding.

Conflicts of Interest: The authors declare no conflict of interest.

References

1.   Ben-Othman, S.; Jõudu, I.; Bhat, R. Bioactives from agri-food wastes: Present insights and future challenges. *Molecules* **2020**, *25*, 510. [CrossRef]

2.   Barbulova, A.; Colucci, G.; Apone, F. New trends in cosmetics: By-products of plant origin and their potential use as cosmetic active ingredients. *Cosmetics* **2015**, *2*, 82–92. [CrossRef]

3.   Schieber, A. Side streams of plant food processing as a source of valuable compounds: Selected examples. *An. Rev. Food Sci. Technol.* **2017**, *8*, 97–112. [CrossRef]

4.   De Camargo, A.C.; Schwember, A.R.; Parada, R.; Garcia, S.; Maróstica Júnior, M.R.; Franchin, M.; Regitano-d'Arce, M.A.B.; Shahidi, F. Opinion on the hurdles and potential health benefits in value-added use of plant food processing by-products as sources of phenolic compounds. *Int. J. Mol. Sci.* **2018**, *19*, 3498. [CrossRef]

5.   Akyol, H.; Riciputi, Y.; Capanoglu, E.; Caboni, M.F.; Verardo, V. Phenolic compounds in the potato and its byproducts: An overview. *Int. J. Mol. Sci.* **2016**, *17*, 835. [CrossRef]

6.   Dos Santos, R.G.; Ventura, P.; Bordado, J.C.; Mateus, M.M. Valorizing potato peel waste: An overview of the latest publications. *Rev. Environ. Sci. Bio Technol.* **2016**, *15*, 585–592. [CrossRef]

7.   Torres, M.D.; Domínguez, H. Valorisation of potato wastes. *Int. J. Food Sci. Technol.* **2019**. [CrossRef]

8.   Chemat, F.; Abert Vian, M.; Ravi, H.K.; Khadhraoui, B.; Hilali, S.; Perino, S.; Fabiano Tixier, A.-S. Review of alternative solvents for green extraction of food and natural products: Panorama, principles, applications and prospects. *Molecules* **2019**, *24*, 3007. [CrossRef]

9.   Park, S. Cyclic glucans enhance solubility of bioavailable flavonoids. *Molecules* **2016**, *21*, 1556. [CrossRef]

10. Suvarna, V.; Gujar, P.; Murahari, M. Complexation of phytochemicals with cyclodextrin derivatives—An insight. *Biomed. Pharmacother.* **2017**, *88*, 1122–1144. [CrossRef]

11. Cai, R.; Yuan, Y.; Cui, L.; Wang, Z.; Yue, T. Cyclodextrin-assisted extraction of phenolic compounds: Current research and future prospects. *Trends Food Sci. Technol.* **2018**, *79*, 19–27. [CrossRef]

12. Lakka, A.; Karageorgou, I.; Kaltsa, O.; Batra, G.; Bozinou, E.; Lalas, S.; Makris, D.P. Polyphenol extraction from *Humulus lupulus* (hop) using a neoteric glycerol/L-alanine deep eutectic solvent: Optimisation, kinetics and the effect of ultrasound-assisted pretreatment. *AgriEngineering* **2019**, *1*, 403–417. [CrossRef]

13. Bezerra, M.A.; Santelli, R.E.; Oliveira, E.P.; Villar, L.S.; Escaleira, L.A. Response surface methodology (RSM) as a tool for optimization in analytical chemistry. *Talanta* **2008**, *76*, 965–977. [CrossRef] [PubMed]

14. Alvarez, V.H.; Cahyadi, J.; Xu, D.; Saldaña, M.D. Optimization of phytochemicals production from potato peel using subcritical water: Experimental and dynamic modeling. *J. Supercrit. Fluids* **2014**, *90*, 8–17. [CrossRef]

15. Kumari, B.; Tiwari, B.K.; Hossain, M.B.; Rai, D.K.; Brunton, N.P. Ultrasound-assisted extraction of polyphenols from potato peels: Profiling and kinetic modelling. *Int. J. Food Sci. Technol.* **2017**, *52*, 1432–1439. [CrossRef]

16. Paleologou, I.; Vasiliou, A.; Grigorakis, S.; Makris, D.P. Optimisation of a green ultrasound-assisted extraction process for potato peel (*Solanum tuberosum*) polyphenols using bio-solvents and response surface methodology. *Biomass Convers. Bioref.* **2016**, *6*, 289–299. [CrossRef]

17. Wu, T.; Yan, J.; Liu, R.; Marcone, M.F.; Aisa, H.A.; Tsao, R. Optimization of microwave-assisted extraction of phenolics from potato and its downstream waste using orthogonal array design. *Food Chem.* **2012**, *133*, 1292–1298. [CrossRef]

18. Riciputi, Y.; Diaz-de-Cerio, E.; Akyol, H.; Capanoglu, E.; Cerretani, L.; Caboni, M.F.; Verardo, V. Establishment of ultrasound-assisted extraction of phenolic compounds from industrial potato by-products using response surface methodology. *Food Chem.* **2018**, *269*, 258–263. [CrossRef]

19. Wijngaard, H.H.; Ballay, M.; Brunton, N. The optimisation of extraction of antioxidants from potato peel by pressurised liquids. *Food Chem.* **2012**, *133*, 1123–1130. [CrossRef]

20. Valcarcel, J.; Reilly, K.; Gaffney, M.; O'Brien, N.M. Antioxidant activity, total phenolic and total flavonoid content in sixty varieties of potato (*Solanum tuberosum* L.) grown in Ireland. *Potato Res.* **2015**, *58*, 221–244. [CrossRef]

21. Zhang, M.; Li, J.; Zhang, L.; Chao, J. Preparation and spectral investigation of inclusion complex of caffeic acid with hydroxypropyl-β-cyclodextrin. *Spectrochim. Acta Part A Mol. Biomol. Spectrosc.* **2009**, *71*, 1891–1895. [CrossRef] [PubMed]

22. Aksamija, A.; Polidori, A.; Plasson, R.; Dangles, O.; Tomao, V. The inclusion complex of rosmarinic acid into beta-cyclodextrin: A thermodynamic and structural analysis by NMR and capillary electrophoresis. *Food Chem.* **2016**, *208*, 258–263. [CrossRef] [PubMed]

23. Álvarez-Parrilla, E.; Palos, R.; Rosa, L.A.; Frontana-Uribe, B.A.; González-Aguilar, G.A.; Machi, L.; Ayala-Zavala, J.F. Formation of two 1:1 chlorogenic acid: β-cyclodextrin complexes at pH 5: Spectroscopic, thermodynamic and voltammetric study. *J. Mex. Chem. Soc.* **2010**, *54*, 103–110. [CrossRef]

24. Goldsmith, C.D.; Vuong, Q.V.; Stathopoulos, C.E.; Roach, P.D.; Scarlett, C.J. Optimization of the aqueous extraction of phenolic compounds from olive leaves. *Antioxidants* **2014**, *3*, 700–712. [CrossRef] [PubMed]

25. Xu, H.; Wang, W.; Liu, X.; Yuan, F.; Gao, Y. Antioxidative phenolics obtained from spent coffee grounds (*Coffea arabica* L.) by subcritical water extraction. *Ind. Crops Prod.* **2015**, *76*, 946–954. [CrossRef]

26. Lakka, A.; Grigorakis, S.; Karageorgou, I.; Batra, G.; Kaltsa, O.; Bozinou, E.; Lalas, S.; Makris, D.P. Saffron processing wastes as a bioresource of high-value added compounds: Development of a green extraction process for polyphenol recovery using a natural deep eutectic solvent. *Antioxidants* **2019**, *8*, 586. [CrossRef]

27. Lakka, A.; Grigorakis, S.; Kaltsa, O.; Karageorgou, I.; Batra, G.; Bozinou, E.; Lalas, S.; Makris, D.P. The effect of ultrasonication pretreatment on the production of polyphenol-enriched extracts from *Moringa oleifera* L. (drumstick tree) using a novel bio-based deep eutectic solvent. *Appl. Sci.* **2020**, *10*, 220. [CrossRef]

28. Cacace, J.; Mazza, G. Mass transfer process during extraction of phenolic compounds from milled berries. *J. Food Eng.* **2003**, *59*, 379–389. [CrossRef]

29. Pinelo, M.; Sineiro, J.; Núñez, M.J. Mass transfer during continuous solid–liquid extraction of antioxidants from grape byproducts. *J. Food Eng.* **2006**, *77*, 57–63. [CrossRef]

30. Rakotondramasy-Rabesiaka, L.; Havet, J.-L.; Porte, C.; Fauduet, H. Estimation of effective diffusion and transfer rate during the protopine extraction process from *Fumaria officinalis* L. *Sep. Purif. Technol.* **2010**, *76*, 126–131. [CrossRef]

31. Philippi, K.; Tsamandouras, N.; Grigorakis, S.; Makris, D.P. Ultrasound-assisted green extraction of eggplant peel (*Solanum melongena*) polyphenols using aqueous mixtures of glycerol and ethanol: Optimisation and kinetics. *Environ. Proc.* **2016**, *3*, 369–386. [CrossRef]

32. Pinelo, M.; Zornoza, B.; Meyer, A.S. Selective release of phenols from apple skin: Mass transfer kinetics during solvent and enzyme-assisted extraction. *Sep. Purif. Technol.* **2008**, *63*, 620–627. [CrossRef]

33. Shewale, S.; Rathod, V.K. Extraction of total phenolic content from *Azadirachta indica* or (neem) leaves: Kinetics study. *Prep. Biochem. Biotechnol.* **2018**, *48*, 312–320. [CrossRef] [PubMed]

34. Stefou, I.; Grigorakis, S.; Loupassaki, S.; Makris, D.P. Development of sodium propionate-based deep eutectic solvents for polyphenol extraction from onion solid wastes. *Clean Technol. Environ. Policy* **2019**, *21*, 1563–1574. [CrossRef]

35. Kaltsa, O.; Lakka, A.; Grigorakis, S.; Karageorgou, I.; Batra, G.; Bozinou, E.; Lalas, S.; Makris, D.P. A green extraction process for polyphenols from elderberry (*Sambucus nigra*) flowers using deep eutectic solvent and ultrasound-assisted pretreatment. *Molecules* **2020**, *25*, 921. [CrossRef]

36. Amado, I.R.; Franco, D.; Sánchez, M.; Zapata, C.; Vázquez, J.A. Optimisation of antioxidant extraction from *Solanum tuberosum* potato peel waste by surface response methodology. *Food Chem.* **2014**, *165*, 290–299. [CrossRef]

37. Frontuto, D.; Carullo, D.; Harrison, S.; Brunton, N.; Ferrari, G.; Lyng, J.; Pataro, G. Optimization of pulsed electric fields-assisted extraction of polyphenols from potato peels using response surface methodology. *Food Bioproc. Technol.* **2019**, *12*, 1708–1720. [CrossRef]

38. Dong, L.; Liu, M.; Chen, A.; Wang, Y.; Sun, D. Solubilities of quercetin in three β-cyclodextrin derivative solutions at different temperatures. *J. Mol. Liquids* **2013**, *177*, 204–208. [CrossRef]

39. Zhang, Q.F.; Cheung, H.Y.; Shangguan, X.; Zheng, G. Structure selective complexation of cyclodextrins with five polyphenols investigated by capillary electrokinetic chromatography. *J. Sep. Sci.* **2012**, *35*, 3347–3353. [CrossRef]

40. Cai, C.; Liu, M.; Yan, H.; Zhao, Y.; Shi, Y.; Guo, Q.; Pei, W.; Han, J.; Wang, Z. A combined calorimetric, spectroscopic and molecular dynamic simulation study on the inclusion complexation of (E)-piceatannol with hydroxypropyl-β-cyclodextrin in various alcohol+ water cosolvents. *J. Chem. Thermodyn.* **2019**, *132*, 341–351. [CrossRef]

41. López-Cobo, A.; Gómez-Caravaca, A.M.; Cerretani, L.; Segura-Carretero, A.; Fernández-Gutiérrez, A. Distribution of phenolic compounds and other polar compounds in the tuber of *Solanum tuberosum* L. by HPLC-DAD-q-TOF and study of their antioxidant activity. *J. Food Compos. Anal.* **2014**, *36*, 1–11. [CrossRef]

42. Albishi, T.; John, J.A.; Al-Khalifa, A.S.; Shahidi, F. Phenolic content and antioxidant activities of selected potato varieties and their processing by-products. *J. Funct. Foods* **2013**, *5*, 590–600. [CrossRef]

Article

The Effect of Ultrasonication Pretreatment on the Production of Polyphenol-Enriched Extracts from *Moringa oleifera* L. (Drumstick Tree) Using a Novel Bio-Based Deep Eutectic Solvent

Achillia Lakka [1], Spyros Grigorakis [2], Olga Kaltsa [1], Ioanna Karageorgou [1], Georgia Batra [1], Eleni Bozinou [1], Stavros Lalas [1] and Dimitris P. Makris [1,*]

[1] School of Agricultural Sciences, University of Thessaly, N. Temponera Street, 43100 Karditsa, Greece; ACHLAKKA@uth.gr (A.L.); okaltsa@uth.gr (O.K.); ikarageorgou@uth.gr (I.K.); gbatra@uth.gr (G.B.); empozinou@uth.gr (E.B.); slalas@uth.gr (S.L.)

[2] Food Quality & Chemistry of Natural Products, Mediterranean Agronomic Institute of Chania (M.A.I.Ch.), International Centre for Advanced Mediterranean Agronomic Studies (CIHEAM), P.O. Box 85, 73100 Chania, Greece; grigorakis@maich.gr

* Correspondence: dimitrismakris@uth.gr; Tel.: +30-24410-64792

Received: 3 December 2019; Accepted: 24 December 2019; Published: 27 December 2019

Abstract: *Moringa oleifera* L. leaves are a plant tissue particularly rich in polyphenolic phytochemicals with significant bioactivities, and there has been significant recent interest for the production of extracts enriched in these substances. The current investigation is aimed at establishing a green extraction process, using a novel eco-friendly natural deep eutectic solvent, composed of glycerol and nicotinamide. Furthermore, sample ultrasonication prior to batch stirred-tank extraction was studied to examine its usefulness as a pretreatment step. Optimization of the extraction process through response surface methodology showed that the maximum total polyphenol yield (82.87 ± 4.28 mg gallic acid equivalents g^{-1} dry mass) could be achieved after a 30 min ultrasonication pretreatment, but the difference with the yield obtained from the non-pretreated sample was statistically non-significant ($p < 0.05$). Extraction kinetics revealed that the activation energy for the ultrasonication-pretreated samples was more energy-demanding, a fact attributed to phenomena pertaining to washing of the readily extracted polyphenols during pretreatment. Liquid-chromatography-diode array-mass spectrometry showed that ultrasonication pretreatment may have a limited positive effect on polyphenol extractability, but the overall polyphenolic profile was identical for the ultrasonication-pretreated and non-pretreated samples.

Keywords: deep eutectic solvents; extraction; *Moringa oleifera*; polyphenols; ultrasonication pretreatment

---

1. Introduction

Moringa oleifera Lam. is a cruciferous plant of the Moringaceae family, often called drumstick tree. *M. oleifera* is a popular staple in several parts around the globe, and it is consumed for its high nutritional value, as its leaves are rich in β-carotene, vitamin C, vitamin E, polyphenols, etc. [1,2]. Moreover, *M. oleifera* has been reported to display a broad range of biological functions, such as neuroprotective activity, chemopreventive properties, anti-inflammatory potency, and hepatoprotective activity. As a number of studies have provided strong evidence regarding the therapeutic value of *M. oleifera* for diseases like diabetes, rheumatoid arthritis, atherosclerosis, etc., this plant has attracted significant attention for its pharmacological functions [3]. The pharmacological potency of *M. oleifera* may be closely associated with its polyphenolic composition, which includes a wide range of structures, represented mainly by flavonoids (kaempferol and quercetin glycosides) and phenolic acids and their derivatives (i.e., caffeoylquinic acids and their isomers) [4,5]. In fact, the strong antioxidant effects exerted by *M. oleifera* leaf extracts and decoctions have been attributed to their high flavonoid content [6–8]. By virtue of their flavonoid richness, *M. oleifera* leaves have been a subject of recent research on the production of polyphenol-containing extracts, and several methodologies have been used for such a purpose, including pressurized hot water [9] and aqueous two-phase extraction [10], ultrasound-assisted extraction with aqueous ethanol [11], and subcritical ethanol extraction [12]. However, extraction with novel, green solvents, the so-called deep eutectic solvents (DES), have lately been proposed as eco-friendly and particularly effective means of polyphenol recovery from *M. oleifera* leaves [13,14]. DES are neoteric designer liquids, composed of food-grade, low-cost biomaterials, such as polyols, organic acids, amines, organic acid salts, etc. [15]. Their facile and straight-forward synthesis, as well as their unique properties, including low or no toxicity, absence of flammability, tunability, recyclability, and biodegradability, make DES ideal solvents for the development of green extraction processes, and over the last five years, DES have attracted particularly high interest for the polyphenol extraction from a wide number of plant tissues [16].

Ultrasonication is a state-of-the-art technology implicated in sustainable "green" extraction procedures, since numerous studies have demonstrated the efficiency of ultrasounds in accelerating the rate of solid–liquid extractions. By implementing ultrasonication techniques, effective extractions can be accomplished in minutes with high reproducibility, requiring reduced solvent volume, lower energy input and simpler sample manipulation, providing at the same time in some cases higher purity of the final product [17]. On such a ground, this study was performed with the aim at examining the extraction of *M. oleifera* leaf polyphenols and the impact of ultrasonication as a pretreatment step. Response surface was deployed to optimize extraction variables and kinetics was used to obtain a deeper insight into the effect of ultrasonication. A final assessment of the process was conducted through the determination of the ultrasonication-treated and non-treated samples.

2. Materials and Methods

2.1. Chemicals

Glycerol anhydrous (99.5%), sodium carbonate anhydrous (99%), aluminium chloride anhydrous (98%), sodium acetate anhydrous (98.5%), and ascorbic acid (99.5%) were from Penta (Praha, Czechia). Gallic acid hydrate (99%) was from Panreac (Barcelona, Spain). Rutin (quercetin 3-*O*-rutinoside) hydrate (>94%), kaempferol 3-*O*-glucoside, 2,4,6-tris (2-pyridyl)-*s*-triazine (TPTZ) (99%), neochlorogenic acid (≥98%) and Folin-Ciocalteu's phenol reagent were from Sigma-Aldrich (St. Louis, MO, USA). Iron chloride hexahydrate was from Merck (Darmstadt, Germany). Nicotinamide and chlorogenic acid (95%) were from Fluorochem (Hadfield, UK).

2.2. Plant Material

M. oleifera trees were cultivated in the Agioi Apostoloi area (at 3922039.400 N and 2154000.600 E and elevation of 100 m, according to Google Earth Pro version 7.3.2.5491 (64 bit) (Google, Inc., Mountain View, CA, USA) of Karditsa prefecture (Thessaly, Greece). The leaves from the plants were collected on the 25th October 2018, when the trees were 2 months old and they were transferred to the laboratory 30 min after collection. Afterwards, they were thoroughly washed with tap water and freeze dried using a Telstar Cryodos 80 freeze dryer (Telstar Industrial, S.A., Terrassa, Spain) for 12 h (the moisture content of the leaves was 8%) and ground in a ball-mill to give a powder with average particle diameter of 0.161 mm. The material was stored in air-tight plastic bags, at −40 °C.

2.3. Synthesis of DES

The procedure followed for the synthesis of all DES used was based on a previous report [18]. Accurately weighted appropriate amounts of glycerol (hydrogen bond donor) and nicotinamide (hydrogen bond acceptor) were mixed in a 100-mL round-bottomed flask and heated up to 80 °C, under stirring at 500 rpm, for an approximate time period of 60 min. This time was sufficient to observe the formation of a transparent liquid. The liquids formed were allowed to cool down to room temperature and stored in sealed glass vials. The vials were checked periodically for the appearance of crystals, to ensure solvent stability. Checking was performed over a period of several weeks.

2.4. Screening Batch Stirred-Tank Extractions

An amount of 0.570 g of dried plant material was transferred into a 50-mL round-bottomed flask with 20 mL of solvent to give a liquid-to-solid ratio ($R_{L/S}$) of 35 mL g^{-1}. The mixture was extracted at 50 °C for 150 min, under continuous stirring at 500 rpm, on a thermostated hot plate (Witeg, Wertheim, Germany). An aliquot of 1 mL of the extract obtained was placed in a 1.5-mL Eppendorf tube and centrifuged at 10,000× g for 10 min. The clear supernatant was used for all analyses. Control extractions were performed with deionised water, 60% (*v/v*) aqueous methanol and 60% (*v/v*) aqueous ethanol, under identical conditions.

2.5. Ultrasound-Assisted Pretreatment

The pretreatment was carried out by ultrasonicating the solvent/solid material mixture before the batch stirred-tank extractions. Ultrasonication was conducted in an Elma D-78224 Singen HTW heated ultrasonic bath (Elma Schmidbauer GmbH, Singen, Germany), operated at a frequency of 50 Hz, a power of 550 W and an acoustic energy density of 78.6 W L^{-1} [19]. Pretreatments were performed at ambient temperature (23 ± 2 °C).

2.6. Response Surface Optimisation of the Extraction

The design of experiment was implemented with the aim at assessing the effect of three selected parameters (independent variables), the DES concentration (C_{DES}, termed as X_1), the liquid-to-solid ratio ($R_{L/S}$, termed as X_2), and the stirring speed (S_S, termed as X_3), on the extraction yield of total polyphenols (Y_{TP}), which represented the response [19]. The mode chosen was Box–Behnken with three central points and the independent variables were coded between −1 (lower limit) and +1 (upper limit), as follows:

$$x_i = \frac{X_i - X_0}{\Delta X_i}, \ i = 1,\ 2,\ 3. \tag{1}$$

The term x_i represents the dimensionless value of the independent variable i and X_i the actual one. X_0 is termed the actual value of the independent variable i at the central point of the design, and ΔX_i the step change of X_i (Table 1). The range of values for the variables used was decided on the basis of previously published results [13,14]. Model significance, as well as the significance for each polynomial coefficient and the overall coefficient R^2 for the mathematical model were evaluated by performing ANOVA, at least at 95% significance level. Non-significant dependent terms ($p > 0.05$) of the model were not considered and were omitted. Model validation was accomplished by runs performed under the predicted optimal conditions. The actual (measured) values were then compared with the predicted ones.

Table 1. Process (independent) variables used for the experimental design and their codified levels.

Independent Variables	Code Units	Coded Variable Level		
		−1	**0**	**1**
C_{DES} (%, *w/v*)	X_1	55	70	85
$R_{L/S}$ (mL g^{-1})	X_2	20	60	100
S_S (rpm)	X_3	200	500	800

2.7. Total Polyphenol (TP) Determination

All samples were diluted 1:50 with 0.5% aqueous formic acid prior to determinations. For TP determination, a protocol described elsewhere was used [20]. Aliquot of 0.1 mL of diluted sample was mixed with 0.1 mL Folin-Ciocalteu reagent in a 1.5-mL Eppendorf tube and allowed to stand for 2 min. Following this, 0.8 mL of Na_2CO_3 solution (5% *w/v*) was added to the mixture and the tube was incubated in a thermostated water bath, at 40 °C, for 20 min. After incubation, the tube was cooled with tap water and the absorbance was recorded at 740 nm, using appropriate blank. Determination of total polyphenol concentration (C_{TP}) was carried out using a gallic acid calibration curve (10–80 mg L^{-1}) and results were expressed as mg gallic acid equivalents (GAE) L^{-1}. The extraction yield in total polyphenols was then calculated as follows:

$$Y_{TP} \ (\text{mg GAE g}^{-1}) = \frac{C_{TP} \times V}{dm} \tag{2}$$

where V is the extraction volume (in L) and dm is the dry mass of the solid material (in g).

2.8. Total Flavonoid (TFn) Determination

The determination was carried out according to a published protocol [21]. Samples were diluted as described for TP analysis and then 0.1 mL sample was mixed with 0.86 mL 35% (*v/v*) aqueous ethanol and 0.04 mL of reagent composed of 5% (*w/v*) $AlCl_3$ and 0.5 M CH_3COONa. The mixture was left to react at room temperature for 30 min, and the absorbance was recorded at 415 nm. Total flavonoid concentration (C_{TFn}) was determined using a calibration curve constructed with rutin as standard (15–300 mg L^{-1}). Results were reported as mg rutin equivalents (RtE) L^{-1}. Yield in TFn (Y_{TFn}) was estimated as described for TP.

2.9. Determination of the Antiradical Activity (A_{AR})

A previously established methodology was employed [22], using DPPH as the radical probe. Each extract was diluted 1:50 prior to analysis, as described for TP determination, and then 0.025 mL of the diluted sample was mixed with 0.975 mL DPPH (100 μM in methanol) and incubated at ambient temperature. The absorbance at 515 nm was read at $t = 0$ min (immediately after mixing) and at $t = 30$ min. The A_{AR} of the extract was determined using the equation:

$$A_{AR} = \frac{C_{DPPH}}{C_{TP}} \times \left(1 - \frac{A_{515(f)}}{A_{515(i)}}\right) \times Y_{TP} \tag{3}$$

where C_{DPPH} and C_{TP} is the DPPH concentration (μM) and total polyphenol concentration (mg L^{-1}) in the reaction mixture, respectively, $A_{515(f)}$ is the A_{515} at $t = 30$ min, and $A_{515(i)}$ is the A_{515} at $t = 0$. Y_{TP} represents the extraction yield (mg g^{-1}) in TP of each extract assayed. A_{AR} was given as μmol DPPH g^{-1} dm.

2.10. Determination of the Reducing Power (P_R)

The ferric-reducing power of the extracts was determined as reported elsewhere [22]. Before the assay, all extracts were diluted 1:50, as described for TP determination. Following this, 0.05 mL sample was incubated with 0.05 mL $FeCl_3$ (4 mM in 0.05 M HCl), for 30 min, at 37 °C, in a water bath. After the completion of the incubation, samples were cooled with tap water and 0.9 mL TPTZ solution (1 mM in 0.05 M HCl) was added. The mixture was left under ambient temperature for further 5 min for colour development and then the absorbance was obtained at 620 nm. P_R was reported as μmol ascorbic acid equivalents (AAE) g^{-1} dm, using an ascorbic acid calibration curve (50–300 μM).

2.11. Liquid Chromatography-Mass Spectrometry (LC-MS)

A modification of a method reported previously was used [23]. The chromatograph was a Finnigan (San Jose, CA, USA) MAT Spectra System P4000 pump, a UV6000LP diode array detector, and a Finnigan AQA mass spectrometer. The column used was a Fortis RP-18 column, 150 mm × 2.1 mm, 3 μm, operated at 40 °C, with a 10-μL injection loop. Mass spectra acquisition with electrospray ionization (ESI) in positive ion mode was accomplished with probe temperature set at 250 °C, acquisition at 10 and 50 eV, detector voltage at 650 V, capillary voltage at 4 kVm and source voltage at 25 V. The eluents used were 2% acetic acid (A) and methanol (B). The flow rate was 0.3 mL min^{-1}, and the elution program used was 0–30 min, 0%–100% methanol, 30–40 min, 100% methanol.

2.12. High-Performance Liquid Chromatography (HPLC) Analyses

A Shimadzu CBM-20A liquid chromatograph (Shimadzu Europa GmbH, Nordrhein-Westfalen, Germany) was used, coupled with a SIL-20AC auto sampler and a Shimadzu SPD-M20A detector. The system was interfaced by Shimadzu LC solution software. Analyses were performed on a Phenomenex Luna C18(2) column (100 Å, 5 μm, 4.6 × 250 mm) (Phenomenex, Inc., Torrance, CA, USA), maintained at a temperature of 40 °C, using 20-μL sample injections. Eluents were (A) 0.5% aqueous formic acid and (B) 0.5% formic acid in MeCN/water (6:4), and the flow rate was 1 mL min^{-1}. The elution program was as follows: 100% A to 60% A in 40 min, 60% A to 50% A in 10 min, 50% A to 30% A in 10 min, and then isocratic elution for another 10 min. Quantification was done at 325 nm (chlorogenates) and 355 nm (flavonols), using calibration curves constructed with chlorogenic acid (1–50 μg mL^{-1}, R^2 = 0.9999), neochlorogenic acid (1–50 μg mL^{-1}, R^2 = 0.9997) kaempferol 3-*O*-glucoside (1–50 μg mL^{-1}, R^2 = 0.9996), and quercetin 3-*O*-rutinoside (1–50 μg mL^{-1}, R^2 = 0.9990).

2.13. Statistical Analyses

Extractions were repeated at least twice, and determinations were carried out at least in triplicate. Values reported are means ± standard deviation (sd). Correlations were performed with regression analysis, at least at a 95% significance level ($p < 0.05$), using SigmaPlot™ 12.5. Experimental design and response surface methodology, as well as every associated statistical analysis (e.g., ANOVA) was performed with JMP™ Pro 13.

3. Results and Discussion

3.1. DES Synthesis and Examination of HBD: HBA Molar Ratio ($R_{mol}^{D/A}$)

The use of nicotinamide as HBA in L-lactic acid-based DES has been recently reported for the first time [24]; however, the use of glycerol as HBD in combination with nicotinamide as HBA, to the best of the authors' knowledge, is heretofore unreported. When designing a DES, the role of $R_{mol}^{D/A}$ in the polyphenol extraction performance is salient [19,25], and therefore, an initial screening of DES with varying $R_{mol}^{D/A}$ was deemed essential in identifying the highest-performing combination.

Attempts to combine glycerol and nicotinamide at $R_{mol}^{D/A}$ up to 4 were not met with success, since the DES synthesized tended to develop crystals within 24 h, at room temperature (23 ± 2 °C). Combination at $R_{mol}^{D/A}$ = 5 produced a DES, which remained stable for several weeks, and based on this finding, a series of DES with $R_{mol}^{D/A}$ ranging from 5 to 13 were synthesized and tested for their effectiveness in extracting polyphenols from *M. oleifera*. All the DES assayed were used as 70% (*w/v*) aqueous mixtures and the outcome of this screening is given in Figure 1. The DES with $R_{mol}^{D/A}$ = 5 exhibited statistically higher extraction efficiency ($p < 0.05$) compared with all other DES and also 60% (*v/v*) aqueous ethanol and 60% (*v/v*) aqueous methanol. Water, on the other hand, was the least efficient solvent ($p < 0.05$). On this ground, the DES with $R_{mol}^{D/A}$ = 5, assigned as GL-Nic (5:1), was chosen for further examination.

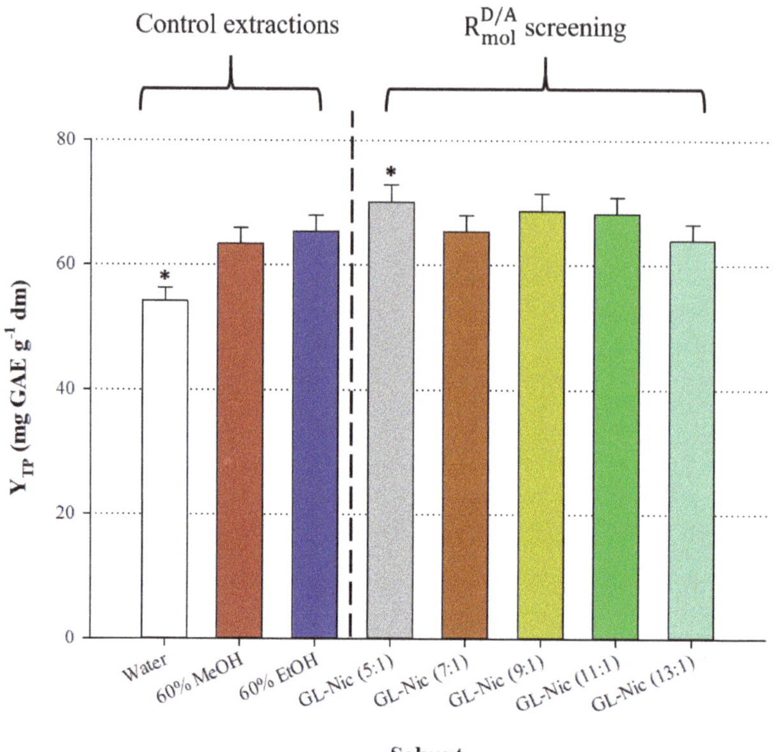

Figure 1. Plot showing the effect of $R_{mol}^{D/A}$ on the extraction effectiveness of the DES tested. Extractions were carried out at 50 °C, with C_{DES} = 70% (*w/v*), at 500 rpm, for 150 min. Assignment with "*" denotes statistically different value.

3.2. The Effect of Ultrasonication Pretreatment

To examine the effect of ultrasonication, GL-Nic (5:1) used as 70% (*w/v*) aqueous solution, was mixed with dried powder of *M. oleifera* leaves, shaken vigorously for a few seconds to form slurry, and then placed in the ultrasonication bath, at room temperature. Room temperature (23 ± 2 °C) was preferred, because at lower temperatures, ultrasonication results in better cavitation, which is the main effect involved in the ultrasound-assisted extraction. At higher temperatures, bubbles formed through cavitation collapse less violently, and thus the ultrasonication effect is less effective [26]. Ultrasonication treatments were performed for 5, 10, 20, 30, and 40 min, followed by stirred-tank extraction. As can be seen in Figure 2, a regime of 30-min ultrasonication gave statistically higher Y_{TP} ($p < 0.05$), whereas extraction without ultrasonication pretreatment (0 min) was significantly less effective ($p < 0.05$). Ultrasonication beyond 30 min resulted in decreased Y_{TP}, thus 30 min was chosen as the most appropriate pretreatment period.

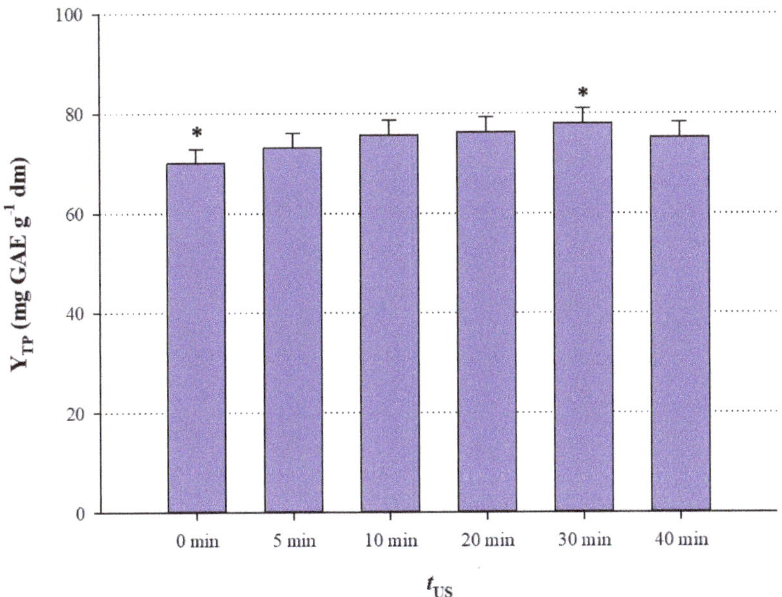

Figure 2. The effect of ultrasonication time (t_{US}) on the extraction yield of total polyphenols. Extractions were performed with the 70% (*w/v*) aqueous GL-Nic (5:1), at 23 °C and $R_{L/S}$ = 35 mL g^{-1} dm. Assignment with "*" denotes statistically different value.

3.3. Optimisation of Extraction Performance

The evidence emerged from recent examinations on polyphenol extraction with DES suggested that the three process (independent) variables considered, namely C_{DES}, $R_{L/S}$, and S_S, are highly influential in solid/liquid extraction processes [19,25]. For this reason, these three parameters were used among several others to build a predictive extraction model. Two cases were examined, one with a 30-min ultrasonication regime prior to every extraction and one without ultrasonication. The purpose of such investigation was to clarify whether an ultrasonication step before batch stirred-tank extraction could affect the optimization settings and assist in achieving significantly higher extraction yield.

Appraisal of the fitted models and response surface suitability were based on the ANOVA test and lack-of-fit test. The p-value for each equation term was calculated to examine the contribution of linear, interaction, and quadratic effects in the independent variables (Table 2). Statistically non-significant terms were omitted from the mathematical models, which are given as polynomial equations in Table 3. The predicted response values calculated by the models, along with the measured (actual) values for each design point, are shown in Table 4. R^2 provides indication of the amount of total variability around the mean explained by the regression model. R^2 terms for both models were ≥0.95, so it would be argued that the estimation of the regression equations exhibited a good adjustment to the experimental data. The p value for lack-of-fit (assuming a confidence interval of 95%) was >0.05 for both models (Table 4), suggesting that the fitted models may be reliable predictors. By setting the values of all three process variables at their optima, using the desirability function (Figure 3), it was made possible to calculate the maximum predicted responses (Table 5). The response surface plots (Figure 4) were helpful in visualizing which variables were more influential to the response.

Table 2. Statistical data obtained after implementing response surface methodology.

Term	Standard Error	t Ratio	p-Value	Sum of Squares	F Ratio
No pretreatment					
C_{DES}	0.778011	5.09	0.0038 *	125.215	25.858
$R_{L/S}$	0.778011	0.72	0.5057	2.48645	0.5135
S_S	0.778011	6.95	0.0009 *	233.820	48.286
$C_{DES}\,R_{L/S}$	1.100274	−3.54	0.0166 *	60.6841	12.531
$C_{DES}\,S_S$	1.100274	−1.33	0.2398	8.61423	1.7789
$R_{L/S}\,S_S$	1.100274	0.40	0.7089	0.75690	0.1563
$C_{DES}C_{DES}$	1.145201	−2.27	0.0726	24.9200	5.1462
$R_{L/S}\,R_{L/S}$	1.145201	0.17	0.8685	0.14708	0.0304
$S_S\,S_S$	1.145201	−0.39	0.7118	0.74079	0.1530
Lack-of-fit			0.1133	22.3468	7.9870
Ultrasonication pretreatment					
C_{DES}	1.074859	63.28	<0.0001 *	28.38811	8.1905
$R_{L/S}$	0.658214	−2.86	0.0353 *	199.7001	57.617
S_S	0.658214	7.59	0.0006 *	21.38580	6.1702
$C_{DES}\,R_{L/S}$	0.658214	−2.48	0.0556	18.19023	5.2482
$C_{DES}\,S_S$	0.930855	2.29	0.0706	18.14760	5.2359
$R_{L/S}\,S_S$	0.930855	−2.29	0.0708	22.46760	6.4824
$C_{DES}C_{DES}$	0.930855	−2.55	0.0515	43.53467	12.560
$R_{L/S}\,R_{L/S}$	0.968865	−3.54	0.0165 *	24.50608	7.0705
$S_S\,S_S$	0.968865	2.66	0.0449 *	30.17521	8.7061
Lack-of-fit			0.4891	11.0748	1.1804

* Asterisk denotes statistically significant values.

Table 3. Mathematical models (equations) derived after implementing response surface methodology (Box–Behnken design).

Case	Polynomial Equations	R^2	p
No pretreatment	$72.98 + 3.96X_1 + 5.41X_3 - 3.90X_1X_2$	0.95	0.0093
Ultrasonication pretreatment	$68.02 - 1.88X_1 + 5.00X_2 - 1.64X_3 + 2.58X_2^2 + 2.88X_3^2$	0.96	0.0055

Table 4. Points of the experimental design considered, and measured and predicted values of the response.

Design Point	Independent Variables			Response (Y_{TP}, mg GAE g^{-1} dm)			
	X_1 (C_{DES}, % w/v)	X_2 ($R_{L/S}$, mL g^{-1})	X_3 (S_S, rpm)	Without UP		With UP	
				Measured	Predicted	Measured	Predicted
1	−1 (55)	−1 (20)	0 (500)	62.82	62.18	66.09	66.18
2	−1 (55)	1 (100)	0 (500)	72.96	71.08	72.58	71.91
3	1 (85)	−1 (20)	0 (500)	76.00	77.88	57.48	58.15
4	1 (85)	1 (100)	0 (500)	70.56	71.20	72.50	72.41
5	0 (70)	−1 (20)	−1 (200)	68.73	67.21	67.03	67.72
6	0 (70)	−1 (20)	1 (800)	76.86	77.15	70.65	69.19
7	0 (70)	1 (100)	−1 (200)	67.74	67.45	81.00	82.46
8	0 (70)	1 (100)	1 (800)	77.61	79.13	75.14	74.45
9	−1 (55)	0 (60)	−1 (200)	56.94	59.11	69.62	68.83
10	1 (85)	0 (60)	−1 (200)	70.31	69.96	70.69	69.33
11	−1 (55)	0 (60)	1 (800)	72.50	72.85	68.46	69.82
12	1 (85)	0 (60)	1 (800)	80.00	77.83	61.01	61.80
13	0 (70)	0 (60)	0 (500)	72.94	72.98	66.07	68.02
14	0 (70)	0 (60)	0 (500)	73.97	72.98	68.47	68.02
15	0 (70)	0 (60)	0 (500)	72.04	72.98	69.52	68.02

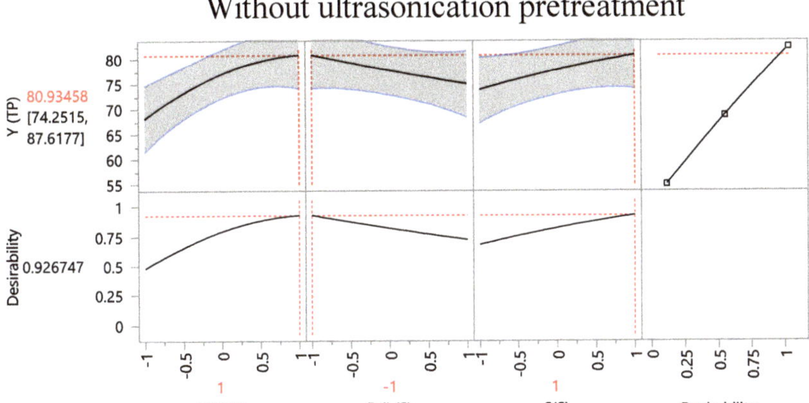

Figure 3. Desirability graphs displaying optimal settings and maximum theoretical extraction yields for total polyphenols (mg GAE g^{-1} dm).

Table 5. Optimal predicted conditions and maximum predicted values (±sd) for the extraction of *M. oleifera* with and without ultrasonication pretreatment.

Case	Maximum Predicted Response (Y_{TP}, mg GAE g^{-1} dm)	Optimal Conditions		
		C_{DES} (*w/v*, %)	$R_{L/S}$ (mL g^{-1})	S_S (rpm)
No pretreatment	80.93 ± 6.68	85	20	800
Ultrasonication pretreatment	82.87 ± 4.28	75	100	200

Ultrasonication of samples prior to stirred-tank extraction had a very pronounced effect on the extraction models, as can be concluded by the equations derived (Table 3). Thus, while C_{DES} (X_1) had a positive contribution in maximizing Y_{TP} in the extraction without ultrasonication pretreatment, it negatively affected the ultrasound-pretreated extraction. On the contrary, increased S_S (X_3) was significant for maximizing Y_{TP} in samples received no pretreatment, whereas ultrasound-pretreated samples required the lowest S_S used to give maximum Y_{TP}.

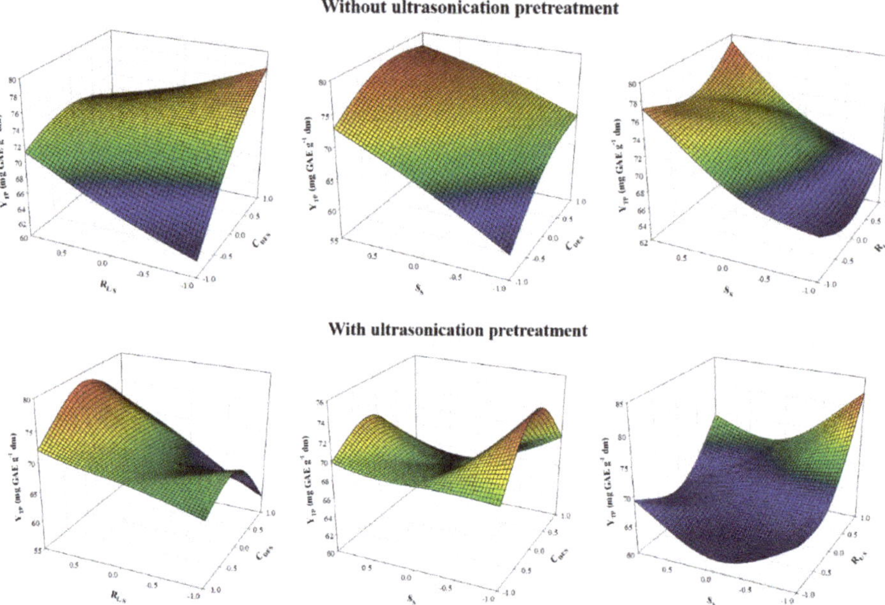

Figure 4. Three-dimensional diagrams given on a comparative arrangement, to illustrate the effect of ultrasonication pretreatment on the variations in the response (Y_{TP}), as a function of different variable levels.

Furthermore, for the extractions without pretreatment, $R_{L/S}$ (X_2) had no direct impact on Y_{TP}, but its cross effect (X_1X_2) with C_{DES} was negative. For the ultrasound-pretreated extractions, quadratic effects of both $R_{L/S}$ (X_2) and S_S (X_2) were positive and significant.

The impact of the ultrasonication pretreatment could be more characteristically depicted by the changes found in the optimization settings of all three variables. Pretreatment resulted in shifting the theoretically required DES amount from 85% to 75% (*w/v*), $R_{L/S}$ from 20 to 100 mL g^{-1}, and S_S from 800 to 200 rpm (Table 5). This outcome suggested a profound influence of ultrasonication on the pattern through which the process variables can affect polyphenol extraction. On the other hand, it is to be emphasized that the ultrasonication pretreatment exerted virtually no effect on the extraction performance, since the optimal predicted values for Y_{TP}, for the extraction without and with pretreatment, had no statistically significant difference, being 80.93 ± 6.68 and 82.87 ± 4.28 mg GAE g^{-1} dm, respectively (Table 5).

A critical assessment of these data would point out that samples received no ultrasonication pretreatment required higher DES concentration and a relatively high speed of stirring to yield extracts with increased Y_{TP}. The optimum C_{DES} determined was 85% (*w/v*), very close to the 80% (*w/v*) found for *M. oleifera* polyphenol extraction with a DES composed of glycerol and sodium acetate (6:1) [13]. Likewise, the optimum S_S (800 rpm) was comparable to 900 rpm determined for polyphenol extraction from onion solid wastes, using a DES of glycerol/sodium propionate (8:1) [25]. By contrast, the optimum $R_{L/S}$ (20 mL g^{-1}) was rather low compared with those reported for polyphenol extraction with DES, ranging from 29 mL g^{-1} [27] to 100 mL g^{-1} [25].

This picture was fundamentally changed when samples were ultrasonicated prior to stirred-tank extraction. The optimum C_{DES} dropped to 75% (*w/v*), while ultrasonication-pretreated samples required 5 times higher $R_{L/S}$ but 4 times lower S_S to provide maximum Y_{TP} (Table 5). Ultrasonication is known to contribute in intensification of extraction efficacy, owed to several phenomena that accompany irradiation with ultrasounds, including propagation of ultrasound pressure waves through the solvent and resulting cavitation effects [28,29]. The increased performance usually observed in the ultrasound-assisted extractions is generally attributed to mechanical, cavitation, and thermal effects, which can provoke cell wall disruption and reduction of particle size, resulting eventually in enhanced mass transfer across cell membranes. On such a theoretical background, it could be postulated that the reduced C_{DES} required for the ultrasonicated-pretreated samples to reach optimum Y_{TP}, could probably be ascribed to higher polyphenol diffusivity and solubilization, as a result of cell wall/membrane disruption. Such an event could facilitate leaching and dissolution of polyphenols into the liquid phase, an assumption corroborated by the drastic decrease in the optimum S_S required, from 800 to 200 rpm. It would appear that lower S_S (200 rpm) is sufficient to provide the turbulence necessary for effective diffusion, whereas higher S_S (800 rpm) might provoke adsorption effects, thereby hindering higher extraction yields.

On the other hand, the pronounced increase in the optimum $R_{L/S}$ could be attributed to a rapid accumulation of polyphenols at the surface of the solid particles, due to cell breakdown and particle disintegration, that would cause extensive liberation of polyphenols from the interior of the solid particles. Such accumulation would presumably require a higher concentration gradient for effective diffusivity, hence the higher $R_{L/S}$. The parameter $R_{L/S}$ is tightly associated with diffusion phenomena [30–32], and it has been demonstrated that rising $R_{L/S}$ could bring about a significant diffusivity increase [33].

3.4. Extraction Kinetics and Temperature Effects

The optimization of the extraction process revealed significant changes in the extraction variables as a consequence of ultrasonication pretreatment, but to portray the ultrasound effect on the process in an integrated frame, information pertaining to the kinetics of the extraction and the effect of temperature was indispensable. Towards this objective, extraction kinetics were traced over a wide breadth of temperatures, ranging from 40 to 80 °C (Figure 5). In each case, the optimized conditions were used to carry out the extractions (Table 5). A modification of a previously proposed kinetic model was employed for both non-pretreated and ultrasound-pretreated samples [34]:

$$Y_{TP(t)} = Y_{TP(0)} + \frac{Y_{TP(s)}t}{t_{0.5} + t}. \tag{4}$$

The term $Y_{TP(t)}$ represents the extraction yield in total polyphenols (mg GAE g^{-1} dm) at any time t, $Y_{TP(0)}$ is a fitting parameter, $Y_{TP(s)}$ the yield in total polyphenols at equilibrium (saturation), and $t_{0.5}$ half the time (min) required for the extraction to enter the regular regime. The phase of the regular regime refers to the period within which small increases in Y_{TP} are achieved within relatively large t [35]. In all cases examined, fitting of the kinetic model to the experimental data gave R^2 > 0.98 (p < 0.0001), suggesting that the model implemented could very effectively describe extraction kinetics. Determination of the initial rate of the extraction, h, and the second-order extraction rate, k, was accomplished using the following equations:

$$h = \frac{Y_{TP(s)}}{t_{0.5}}, \tag{5}$$

$$k = \frac{1}{Y_{TP(s)}\, t_{0.5}}. \tag{6}$$

The data generated from the kinetic assay are analytically presented in Table 6. The effect exerted by the ultrasonication pretreatment was shown to be temperature-dependent, since acceleration of the extraction (increased k) was seen only at 70 and 80 °C. This trend was corroborated by the values determined for h, which were increased for the ultrasound-pretreated samples at temperatures higher than 60 °C. Likewise, $t_{0.5}$ was shorter for the extractions received ultrasonication pretreatment at temperatures higher than 60 °C. Furthermore, for every temperature tested, Y_{TP} of the ultrasonication-pretreated samples was always significantly higher than that of the non-pretreated ones. However, in both cases, temperature had a non-significant effect on $Y_{TP(s)}$, since it varied between 82.03 to 83.90 mg GAE g^{-1} dm for the non-pretreated extractions and from 91.24 to 93.97 mg GAE g^{-1} dm for the ultrasound-pretreated extractions.

Without ultrasonication pretreatment

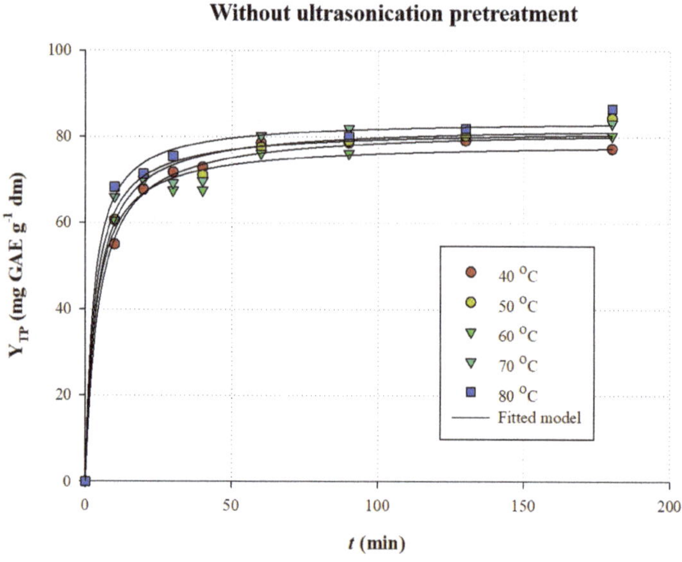

With ultrasonication pretreatment

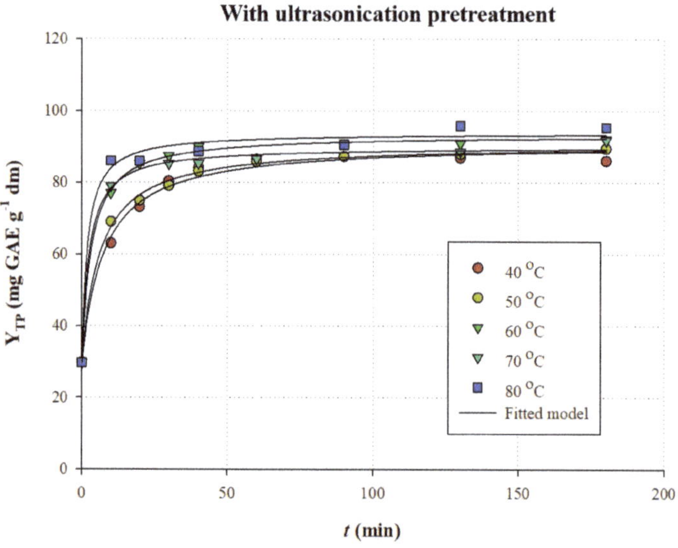

Figure 5. Kinetics of polyphenol extraction from *M. oleifera* leaves, with and without prior ultrasonication pretreatment. Extraction conditions for the non-pretreated sample: $C_{DES} = 85\%$ (*w/w*); $R_{L/S} = 20$ mL g^{-1}; $S_S = 800$ rpm. Extraction conditions for the ultrasonication-pretreated sample: $C_{DES} = 75\%$ (*w/w*); $R_{L/S} = 100$ mL g^{-1}; $S_S = 200$ rpm.

Table 6. Kinetic data and the effect of temperature.

T (°C)	Kinetic Parameters				
	k (×10^{-3}) (g mg^{-1} min^{-1})	h (mg g^{-1} min^{-1})	$Y_{TP(s)}$ (mg GAE g^{-1})	$t_{0.5}$ (min)	E_a (kJ mol^{-1})
No pretreatment					
40	2.68	18.03	82.03	4.55	
50	3.17	21.71	82.71	3.81	
60	3.58	22.15	78.62	3.55	11.24
70	3.95	26.32	81.60	3.10	
80	4.40	30.96	83.90	2.71	
Ultrasonication pretreatment					
40	1.46	12.12	91.24	7.53	
50	1.76	14.58	91.01	6.24	
60	3.29	28.78	93.52	3.25	34.02
70	4.52	36.79	90.14	2.45	
80	5.72	50.52	93.97	1.86	

Pretreatment with ultrasonication is likely to solubilize all polyphenols occurring at and near the surface of the solid particles. This assumption could justify the fact that at temperatures 40–60 °C the stirred-tank extraction of the ultrasound-pretreated samples proceeded at a lower rate. On the other hand, the non-pretreated samples had a higher load of polyphenols at the exterior of the solid particles and exhibited apparently increased k and h. At higher temperatures, polyphenol extraction was more rapid for the ultrasonicated samples, most probably because the pretreatment brought about disruption of the cell walls of the plant material, which enabled more facile penetration of the solvent into the solid particles and easier entrainment of the solute (polyphenols) in the liquid phase. Concerning $Y_{TP(s)}$, the variation as a response to T was non-significant, a fact indicating the T had virtually no effect on $Y_{TP(s)}$. This outcome is a paradox, in light of previous investigations, which demonstrated that increasing T had a proportional effect on $Y_{TP(s)}$ [36,37]. However, earlier studies on polyphenol extraction from *M. oleifera* leaves with a glycerol/sodium acetate DES revealed that switching T from 50 to 80 °C resulted in a constant decline in $Y_{TP(s)}$ [13]. The fact that increasing T did not contribute in attaining higher $Y_{TP(s)}$ might suggest that extraction could reach equilibrium after a given period of time, irrespective of the T.

The increase in k as a response to increasing T was found to obey the Arrhenius law:

$$k = k_0 e^{-\frac{E_a}{RT}}. \tag{7}$$

where k_0, R, T, and E_a correspond to the temperature-independent factor (min^{-1}), the universal gas constant (8.314 J K^{-1} mol^{-1}), the absolute temperature (K), and the activation energy (J mol^{-1}). The linear expression of Equation (7) would be:

$$lnk = lnk_0 + \left(-\frac{E_a}{R}\right)\frac{1}{T}. \tag{8}$$

Thus, E_a could be determined graphically, by the slope ($-\frac{E_a}{R}$) of the straight line obtained by plotting lnk as a function of $1/T$ (Figure 6).

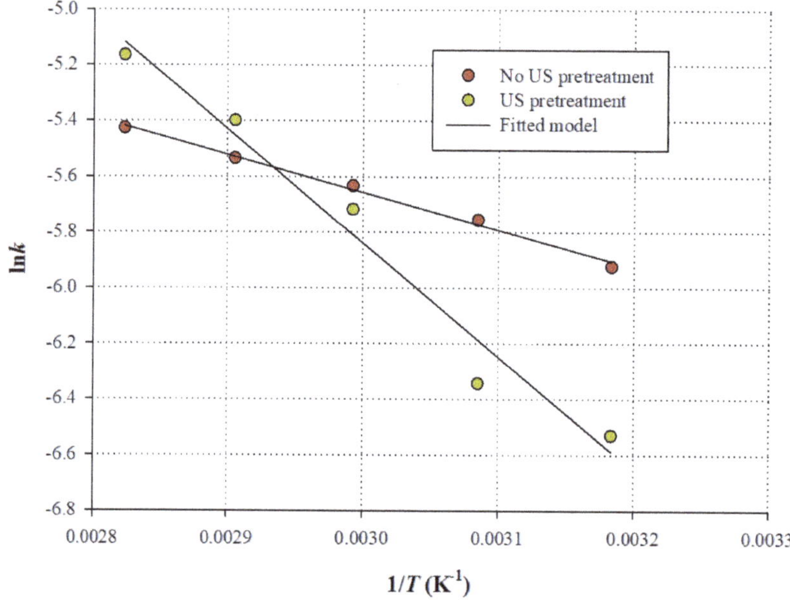

Figure 6. Arrhenius plot for the extractions with no pretreatment (No US) and ultrasonication pretreatment (US).

The E_a values for the extractions without and with ultrasonication pretreatment were 11.24 and 34.02 kJ mol^{-1}, respectively (Table 6), clearly showing that the extraction of the ultrasonication-pretreated samples was more energy-demanding by almost 4 times. Based on this outcome, it would appear that ultrasonication prior to batch stirred-tank extraction does not lower the requirements in energy for accomplishing an efficient extraction of polyphenols from *M. oleifera* leaves, but the energetic needs are even higher.

Yet, such observation could be rather misleading and the increased of $Y_{TP(s)}$ of the ultrasonication-pretreated samples should not be overlooked. The increased E_a value found for the ultrasonication pretreated extraction was most probably because during pretreatment the most readily extracted polyphenols were dissolved in the liquid phase (solvent), leaving behind (inside the solid particles) polyphenols that were more difficult to extract, since solubilization of these polyphenols would depend on internal diffusion, which is the rate-determining step in diffusion-controlled extractions [30,38]. Therefore, the following stirred-tank extraction was apparently more energy-demanding. However, considering $Y_{TP(s)}$, it could be argued that ultrasonication facilitated the washing (initial) phase of the extraction, provoking extended release of superficial polyphenols, whereas the stirred-tank extraction recovered the remaining compounds, located deeper inside the particles. Indeed, just after ultrasonication the Y_{TP} was approximately 29.70 mg GAE g^{-1} dm, which was almost 30% of the $Y_{TP(s)}$ achieved with ultrasonication pretreatment (Table 6). These data strongly supported that a large part of polyphenols was extracted during the ultrasonication step.

3.5. Polyphenolic Profile

To illustrate the effect of ultrasonication pretreatment on the analytical polyphenolic profile, LC-MS was undertaken for the samples obtained after 180 min of extraction, at 80 °C, under optimal conditions. A typical chromatogram is given in Figure 7, showing the principal substances detected at both 320 and 360 nm. It should be noted that for both ultrasonication-pretreated and non-pretreated samples, the chromatograms recorded were identical (data not shown), evidencing no effect of ultrasonication on the polyphenolic profile.

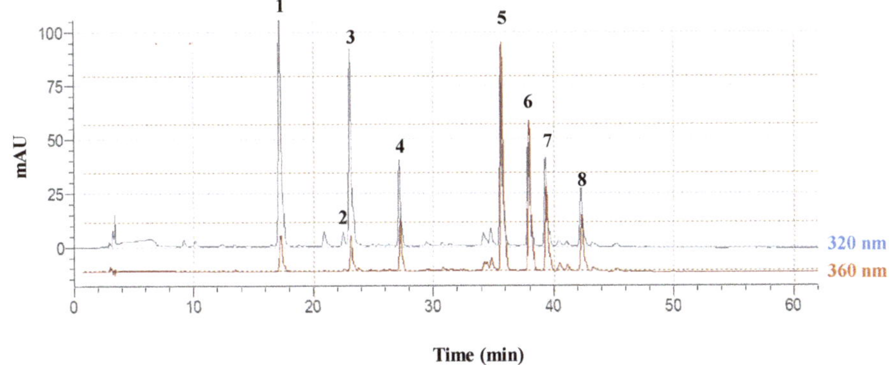

Time (min)

Figure 7. HPLC traces of the sample extracted under optimal conditions, at 80 °C, with ultrasonication pretreatment, recorded at 320 and 360 nm. Peak assignment: 1, neochlorogenic acid; 2, chlorogenic acid; 3, chlorogenic acid isomer; 4, multiflorin B; 5, quercetin glucoside; 6, quercetin manonylglycoside derivative; 7, kaempferol glucoside; 8, kaempferol malonylglucoside.

Peak #1 showed a pseudo-molecular ion at m/z = 355 and a derivative ion at m/z = 163. Considering the UV-vis characteristics, this structure was assigned to neochlorogenic acid. The identity of this peak was further confirmed by comparing the UV-vis spectrum and retention time (Rt) with an original standard. Likewise, peaks 2 and 3 were tentatively identified as chlorogenic acid and a chlorogenate isomer, respectively (Table 7).

Table 7. Analytical spectrometric data of the major polyphenolic phytochemicals detected in the extracts of *M. oleifera* leaves.

No	Rt (min)	UV-Vis	$[M + H]^+$ (m/z)	Fragment Ions (m/z)	Tentative Identity
1	17.24	244, 318	355	163	Neochlorogenic acid
2	22.48	246, 318	355	377, 163	Chlorogenic acid
3	23.07	246, 318	355	377, 163	Chlorogenic acid isomer
4	27.19	270, 320(s), 340	595	617$[M + Na]^+$, 287	Multiflorin B
5	35.62	256, 318(s), 354	465	487$[M + Na]^+$, 303	Quercetin glucoside
6	37.89	256, 316(s), 360	833	609, 573, 551, 303	Quercetin malonylglycoside derivative
7	39.34	264, 318(s), 348	449	471$[M + Na]^+$, 287	Kaempferol glucoside
8	42.30	264, 348	535	557$[M + Na]^+$, 287	Kaempferol malonylglucoside

Peak #4 displayed a pseudo-molecular ion at m/z = 595, a sodium adduct at m/z = 617, and a diagnostic fragment (aglycone) at m/z = 287. This compound was tentatively assigned to multiflorin B (kaempferol 3-*O*-rhamnosylglucoside). Similarly, peaks 7 and 8 were assigned to a kaempferol glucoside and kaempferol malonylglucoside. Peaks 5 and 6, which exhibited typical quercetin glycoside UV-vis pattern and diagnostic fragment at m/z = 303, were tentatively identified as quercetin glucoside and quercetin malonylglycoside derivative, respectively [39]. All these compounds have been previously reported in *M. oleifera* leaf extracts [13]; however, the presence of a quercetin rhamnoside derivative claimed in an earlier study [40] was not confirmed.

The information emerged from the quantitative data (Table 8) showed that ultrasonication pretreatment afforded changes in the recovery of major polyphenols, but not all substances were equally affected.

Table 8. Quantitative data for the major polyphenols detected in *M. oleifera* extracts.

Compound	Content (mg g^{-1} dm) $\pm$ %rsd	
	No Pretreatment	**Ultrasonication Pretreatment**
Neochlorogenic acid	6.61 $\pm$ 0.10	6.82 $\pm$ 0.11
Chlorogenic acid	0.20 $\pm$ 0.00	0.22 $\pm$ 0.00
Chlorogenic acid isomer	5.32 $\pm$ 0.07	5.22 $\pm$ 0.08
Total chlorogenates	*12.13*	*12.26*
Multiflorin B	2.91 $\pm$ 0.03	3.11 $\pm$ 0.03
Quercetin glucoside	10.58 $\pm$ 0.11	10.65 $\pm$ 0.12
Quercetin malonylglycoside derivative	5.97 $\pm$ 0.06	6.99 $\pm$ 0.07
Kaempferol glucoside	3.87 $\pm$ 0.04	4.16 $\pm$ 0.05
Kaempferol malonylglucoside	0.37 $\pm$ 0.00	0.42 $\pm$ 0.00
Total flavonols	*23.70*	*25.33*
Total polyphenols	*35.83*	*37.59*

For neochlorogenic acid and chlorogenic acid, ultrasonication pretreatment brought about an increase by 3.1% and 9.1%, but for the chlorogenic acid isomer, a decrease by almost 1.9% was observed. Overall, total chlorogenate content was by 1.1% increased, a difference that falls within the limits of statistical error. Thus, ultrasonication pretreatment had practically no effect on chlorogenate extractability. On the other hand, all flavonol glycosides had higher content in the ultrasonication-pretreated sample, the increases varying from 0.65% (quercetin glucoside) to 14.6% (quercetin manolyglycoside derivative). Overall, the increase in flavonol content caused by ultrasonication pretreatment was 6.4%, and for all polyphenols considered (chlorogenates + flavonols), an increase by 4.7% was determined.

4. Conclusions

The extraction of *M. oleifera* leaf polyphenols with a novel, glycerol/nicotinamide DES was performed by deploying an ultrasonication pretreatment step. This process enabled the clarification of the use of ultrasounds in the performance of the extraction, with regard to yield, duration, and energy demands. The optimization through response surface methodology revealed that when samples were ultrasonication-pretreated, the subsequent batch stirred-tank extraction required less amount of DES and much lower stirring speed for maximum polyphenol yield, but also a higher amount of solvent per dry mass unit. However, no significant difference in the maximum polyphenol yield was determined. Kinetics showed that higher extraction rate, and therefore shorter extraction time, of the ultrasound-pretreated samples, could be attained only when stirred-tank extraction was carried out at $T \geq 70$ °C. The higher energy barrier estimated for the ultrasonication-pretreated extractions was representative of the harder-to-extract compounds, since a significant wash out of the most readily extracted polyphenols took place during pretreatment. It is concluded that sample ultrasonication as a pretreatment step might be favourable in reducing extraction time, and solvent and energy requirements, but further studies with various plant tissues are needed to obtain a deeper insight.

Author Contributions: Conceptualization, S.L. and D.P.M.; formal analysis, A.L., S.G., I.K., O.K., G.B., and E.B.; investigation, A.L., S.G., I.K., O.K., G.B., and E.B.; writing—original draft preparation, S.L. and D.P.M.; writing—review and editing, S.L. and D.P.M.; supervision, S.L. and D.P.M. All authors have read and agreed to the published version of the manuscript.

Funding: This research was co-financed by the European Union and the Hellenic Ministry of Economy & Development through the Operational Programme Competitiveness, Entrepreneurship and Innovation, under the call RESEARCH—CREATE—INNOVATE (project code: T1EDK-05677).

Conflicts of Interest: The authors declare no conflict of interest.

Nomenclature

A_{AR}	antiradical activity (μmol DPPH g^{-1})
dm	dry matter (g)
P_R	reducing power (μmol AAE g^{-1})
$R_{L/S}$	liquid-to-solid ratio (mL g^{-1})
$R_{mol}^{D/A}$	molar HBD:HBA ratio (dimensionless)
S_S	stirring speed (rpm)
t	time (min)
T	temperature ($^\circ$C)
Y_{TP}	yield in total polyphenols (mg GAE g^{-1})

Abbreviations

AAE	ascorbic acid equivalents
DES	deep eutectic solvents
DPPH	2,2-diphenyl-1-picrylhydrazyl radical
GAE	gallic acid equivalents
HBA	hydrogen bond acceptor
HBD	hydrogen bond donor
TPTZ	2,4,6-tripyridyl-*s*-triazine
UP	ultrasonication pretreatment

References

1. Gopalakrishnan, L.; Doriya, K.; Kumar, D.S. *Moringa oleifera*: A review on nutritive importance and its medicinal application. *Food Sci. Hum. Wellness* **2016**, *5*, 49–56. [CrossRef]
2. Kou, X.; Li, B.; Olayanju, J.B.; Drake, J.M.; Chen, N. Nutraceutical or pharmacological potential of *Moringa oleifera* Lam. *Nutrients* **2018**, *10*, 343. [CrossRef] [PubMed]
3. Gupta, S.; Jain, R.; Kachhwaha, S.; Kothari, S. Nutritional and medicinal applications of *Moringa oleifera* Lam.—Review of current status and future possibilities. *J. Herb. Med.* **2018**, *11*, 1–11. [CrossRef]
4. Lin, M.; Zhang, J.; Chen, X. Bioactive flavonoids in *Moringa oleifera* and their health-promoting properties. *J. Funct. Foods* **2018**, *47*, 469–479. [CrossRef]
5. Saucedo-Pompa, S.; Torres-Castillo, J.; Castro-López, C.; Rojas, R.; Sánchez-Alejo, E.; Ngangyo-Heya, M.; Martínez-Ávila, G. *Moringa* plants: Bioactive compounds and promising applications in food products. *Food Res. Int.* **2018**, *111*, 438–450. [CrossRef]
6. Lalas, S.; Athanasiadis, V.; Karageorgou, I.; Batra, G.; Nanos, G.D.; Makris, D.P. Nutritional characterization of leaves and herbal tea of *Moringa oleifera* cultivated in Greece. *J. Herbs Spices Med. Plants* **2017**, *23*, 320–333. [CrossRef]
7. Rani, A.; Zahirah, N.; Husain, K.; Kumolosasi, E. *Moringa* genus: A review of phytochemistry and pharmacology. *Front. Pharmacol.* **2018**, *9*, 108. [CrossRef]
8. Singh, A.K.; Rana, H.K.; Tshabalala, T.; Kumar, R.; Gupta, A.; Ndhlala, A.R.; Pandey, A.K. Phytochemical, nutraceutical and pharmacological attributes of a functional crop *Moringa oleifera* Lam: An overview. *S. Afr. J. Bot.* **2019**, in press. [CrossRef]
9. Nuapia, Y.; Cukrowska, E.; Tutu, H.; Chimuka, L. Statistical comparison of two modeling methods on pressurized hot water extraction of vitamin C and phenolic compounds from *Moringa oleifera* leaves. *S. Afr. J. Bot.* **2018**, in press. [CrossRef]
10. Djande, C.H.; Piater, L.; Steenkamp, P.; Madala, N.; Dubery, I. Differential extraction of phytochemicals from the multipurpose tree, *Moringa oleifera*, using green extraction solvents. *S. Afr. J. Bot.* **2018**, *115*, 81–89. [CrossRef]
11. Hamed, Y.; Abdin, M.; Akhtar, H.; Chen, D.; Wan, P.; Chen, G.; Zeng, X. Extraction, purification by macrospores resin and in vitro antioxidant activity of flavonoids from *Moringa oliefera* leaves. *S. Afr. J. Bot.* **2019**, *124*, 270–279. [CrossRef]

12. Wang, Y.; Gao, Y.; Ding, H.; Liu, S.; Han, X.; Gui, J.; Liu, D. Subcritical ethanol extraction of flavonoids from *Moringa oleifera* leaf and evaluation of antioxidant activity. *Food Chem.* **2017**, *218*, 152–158. [CrossRef] [PubMed]

13. Karageorgou, I.; Grigorakis, S.; Lalas, S.; Makris, D.P. Enhanced extraction of antioxidant polyphenols from *Moringa oleifera* Lam. leaves using a biomolecule-based low-transition temperature mixture. *Eur. Food Res. Technol.* **2017**, *243*, 1839–1848. [CrossRef]

14. Karageorgou, I.; Grigorakis, S.; Lalas, S.; Mourtzinos, I.; Makris, D.P. Incorporation of 2-hydroxypropyl β-cyclodextrin in a biomolecule-based low-transition temperature mixture (LTTM) boosts efficiency of polyphenol extraction from *Moringa oleifera* Lam leaves. *J. Appl. Res. Med. Aromat. Plants* **2018**, *9*, 62–69. [CrossRef]

15. Mišan, A.; Nađpal, J.; Stupar, A.; Pojić, M.; Mandić, A.; Verpoorte, R.; Choi, Y.H. The perspectives of natural deep eutectic solvents in agri-food sector. *Crit. Rev. Food Sci. Nutr.* **2019**, in press.

16. De los Ángeles Fernández, M.; Boiteux, J.; Espino, M.; Gomez, F.V.; Silva, M.F. Natural deep eutectic solvents-mediated extractions: The way forward for sustainable analytical developments. *Anal. Chim. Acta* **2018**, *1038*, 1–10. [CrossRef]

17. Chemat, F.; Rombaut, N.; Sicaire, A.-G.; Meullemiestre, A.; Fabiano-Tixier, A.-S.; Abert-Vian, M. Ultrasound assisted extraction of food and natural products. Mechanisms, techniques, combinations, protocols and applications. A review. *Ultrason. Sonochem.* **2017**, *34*, 540–560. [CrossRef]

18. Mouratoglou, E.; Malliou, V.; Makris, D.P. Novel glycerol-based natural eutectic mixtures and their efficiency in the ultrasound-assisted extraction of antioxidant polyphenols from agri-food waste biomass. *Waste Biomass Valorization* **2016**, *7*, 1377–1387. [CrossRef]

19. Lakka, A.; Karageorgou, I.; Kaltsa, O.; Batra, G.; Bozinou, E.; Lalas, S.; Makris, D. Polyphenol extraction from *Humulus lupulus* (hop) using a neoteric glycerol/L-alanine deep eutectic solvent: Optimisation, kinetics and the effect of ultrasound-assisted pretreatment. *AgriEngineering* **2019**, *1*, 403–417. [CrossRef]

20. Cicco, N.; Lanorte, M.T.; Paraggio, M.; Viggiano, M.; Lattanzio, V. A reproducible, rapid and inexpensive Folin–Ciocalteu micro-method in determining phenolics of plant methanol extracts. *Microchem. J.* **2009**, *91*, 107–110. [CrossRef]

21. Manousaki, A.; Jancheva, M.; Grigorakis, S.; Makris, D. Extraction of antioxidant phenolics from agri-food waste biomass using a newly designed glycerol-based natural low-transition temperature mixture: A comparison with conventional eco-friendly solvents. *Recycling* **2016**, *1*, 194–204. [CrossRef]

22. Paleologou, I.; Vasiliou, A.; Grigorakis, S.; Makris, D.P. Optimisation of a green ultrasound-assisted extraction process for potato peel (*Solanum tuberosum*) polyphenols using bio-solvents and response surface methodology. *Biomass Convers. Biorefin.* **2016**, *6*, 289–299. [CrossRef]

23. Karakashov, B.; Grigorakis, S.; Loupassaki, S.; Makris, D.P. Optimisation of polyphenol extraction from *Hypericum perforatum* (St. John's Wort) using aqueous glycerol and response surface methodology. *J. Appl. Res. Med. Aromat. Plants* **2015**, *2*, 1–8. [CrossRef]

24. Georgantzi, C.; Lioliou, A.-E.; Paterakis, N.; Makris, D. Combination of lactic acid-based deep eutectic solvents (DES) with β-cyclodextrin: Performance screening using ultrasound-assisted extraction of polyphenols from selected native Greek medicinal plants. *Agronomy* **2017**, *7*, 54. [CrossRef]

25. Stefou, I.; Grigorakis, S.; Loupassaki, S.; Makris, D.P. Development of sodium propionate-based deep eutectic solvents for polyphenol extraction from onion solid wastes. *Clean Technol. Environ. Policy* **2019**, *21*, 1563–1574. [CrossRef]

26. Ameta, R. Sonochemistry: A Versatile Approach. In *Sonochemistry*; Apple Academic Press: Palm Bay, FL, USA, 2018; pp. 371–373.

27. Jancheva, M.; Grigorakis, S.; Loupassaki, S.; Makris, D.P. Optimised extraction of antioxidant polyphenols from *Satureja thymbra* using newly designed glycerol-based natural low-transition temperature mixtures (LTTMs). *J. Appl. Res. Med. Aromat. Plants* **2017**, *6*, 31–40. [CrossRef]

28. Shirsath, S.; Sonawane, S.; Gogate, P. Intensification of extraction of natural products using ultrasonic irradiations—A review of current status. *Chem. Eng. Process.* **2012**, *53*, 10–23. [CrossRef]

29. Pingret, D.; Fabiano-Tixier, A.S.; Chemat, F. Ultrasound-assisted extraction. Natural product extraction: Principles and applications. *RSC Green Chem.* **2013**, *21*, 89–112.

30. Pinelo, M.; Sineiro, J.; Núñez, M.a.J. Mass transfer during continuous solid–liquid extraction of antioxidants from grape byproducts. *J. Food Eng.* **2006**, *77*, 57–63. [CrossRef]

31. Rakotondramasy-Rabesiaka, L.; Havet, J.-L.; Porte, C.; Fauduet, H. Estimation of effective diffusion and transfer rate during the protopine extraction process from *Fumaria officinalis* L. *Sep. Purif. Technol.* **2010**, *76*, 126–131. [CrossRef]

32. Krishnan, R.Y.; Rajan, K. Microwave assisted extraction of flavonoids from *Terminalia bellerica*: Study of kinetics and thermodynamics. *Sep. Purif. Technol.* **2016**, *157*, 169–178. [CrossRef]

33. Shewale, S.; Rathod, V.K. Extraction of total phenolic content from *Azadirachta indica* or (neem) leaves: Kinetics study. *Prep. Biochem. Biotechnol.* **2018**, *48*, 312–320. [CrossRef] [PubMed]

34. Makris, D. A novel kinetic assay for the examination of solid-liquid extraction of flavonoids from plant material. *Res. J. Chem. Sci.* **2015**, *5*, 18–23.

35. Seikova, I.; Simeonov, E.; Ivanova, E. Protein leaching from tomato seed—experimental kinetics and prediction of effective diffusivity. *J. Food Eng.* **2004**, *61*, 165–171. [CrossRef]

36. Trasanidou, D.; Apostolakis, A.; Makris, D.P. Development of a green process for the preparation of antioxidant and pigment-enriched extracts from winery solid wastes using response surface methodology and kinetics. *Chem. Eng. Commun.* **2016**, *203*, 1317–1325. [CrossRef]

37. Athanasiadis, V.; Grigorakis, S.; Lalas, S.; Makris, D.P. Highly efficient extraction of antioxidant polyphenols from *Olea europaea* leaves using an eco-friendly glycerol/glycine deep eutectic solvent. *Waste Biomass Valorization* **2018**, *9*, 1985–1992. [CrossRef]

38. Karacabey, E.; Mazza, G.; Bayındırlı, L.; Artık, N. Extraction of bioactive compounds from milled grape canes (*Vitis vinifera*) using a pressurized low-polarity water extractor. *Food Bioprocess Technol.* **2012**, *5*, 359–371. [CrossRef]

39. Lin, H.; Zhu, H.; Tan, J.; Wang, H.; Wang, Z.; Li, P.; Zhao, C.; Liu, J. Comparative analysis of chemical constituents of *Moringa oleifera* leaves from China and India by ultra-performance liquid chromatography coupled with quadrupole-time-of-flight mass spectrometry. *Molecules* **2019**, *24*, 942. [CrossRef]

40. Karageorgou, I.; Grigorakis, S.; Lalas, S.; Makris, D.P. The effect of 2-hydroxypropyl β-cyclodextrin on the stability of polyphenolic compounds from *Moringa oleifera* Lam leaf extracts in a natural low-transition temperature mixture. *Nova Biotechnol. Chim.* **2018**, *17*, 29–37. [CrossRef]

Article

Application of Deep Eutectic Solvents to Prepare Mixture Extracts of Three Long-Lived Trees with Maximized Skin-Related Bioactivities

Yan Jin [1], Dasom Jung [1], Ke Li [1], Keunbae Park [1], Jaeyoung Ko [2,*], Misuk Yang [2] and Jeongmi Lee [1,*]

[1] School of Pharmacy, Sungkyunkwan University, Suwon, Gyeonggi-do 16419, Korea
[2] Amorepacific Research and Development Center, Giheung-gu, Yongin 17074, Korea
*   Correspondence: jaeyoungko@amorepacific.com (J.K.); jlee0610@skku.edu (J.L.);
    Tel.: +82-31-290-7784 (J.L.); Fax: +82-31-292-8800 (J.L.)

Received: 20 May 2019; Accepted: 22 June 2019; Published: 26 June 2019

Abstract: This study aims to apply deep eutectic solvents (DESs) as safe and efficient extraction media that could yield maximized skin-related bioactivities from a mixture of long-lived trees. *Ginkgo biloba* L., *Cinnamomum camphora* (L.) J. Presl., and *Cryptomeria japonica* (L.f.) D. Don, native to Asia, were examined as potential resources of cosmeceutical products. Various DESs synthesized from cosmetics-compatible compounds were used to prepare leaf extracts. A DES containing glycerol and xylitol yielded the highest extractability for isoquercetin, and was selected as the optimal solvent. Isoquercetin has various bioactivities and was found in the extracts of the leaves of all three trees. Then, a series of mixtures of the tree leaves were prepared according to a simplex-centroid mixture design, and their DES-extracts were tested for skin-related activities, including antioxidant, anti-tyrosinase, and anti-elastase activities. The mixture design resulted in two special cubic models and one quadratic model best fitted for describing the antioxidant and anti-elastase activities, and the anti-tyrosinase activity, respectively. Based on the established models, three different optimal formulations of the three kinds of tree leaves were suggested for maximized responses. The present strategy, which is based on the simplex-centroid mixture design with a DES as the extraction solvent, could be applied to developing new materials from a mixture of natural resources, suitable for the cosmetics and related fields.

Keywords: cosmeceutical product; long-lived trees; simplex-centroid mixture design; deep eutectic solvents; bioactivity

1. Introduction

The skin can be damaged by numerous factors including ultraviolet radiation, smoking, hormonal imbalance, ethanol ingestion, air pollution, and inflammation [1,2]. Cosmetics are most relied upon for skin damage prevention because they are more convenient and economical, and have fewer side effects than pharmaceuticals or cosmetic surgery [3]. Thus, this has increased the demand for effective cosmeceutical products containing natural bioactive compounds [4].

The need for natural bioactive compound-containing cosmeceutical products might be met by a certain category of plants containing compounds with useful bioactivities in the human skin. Asian countries, like Korea, Japan, and China, have always used traditional medicines for various diseases, and have been at the forefront of cosmeceutical product production from natural resources [5]. By actively searching for long-lived trees native to Asia that may contain compounds with skin-beneficial effects, we found *Ginkgo biloba* L. (GB), *Cinnamomum camphora* (L.) J. Presl. (CC), and *Cryptomeria japonica* (L.f.) D. Don (CJ). The living fossil GB has continuously gained popularity as a dietary supplement [6] and complementary medicine [7]. Extracts of its leaves are rich in ginkgolides, flavonoids, biflavones, terpenoids, and polyphenols [8,9], and have been used for treating age-related diseases [10]. GB is also used in anti-aging cosmetics due to its antioxidant, anti-lipid peroxidation, and anti-inflammatory activities [11]. The evergreen tree CC, commonly known as the camphor tree, is used as a medicinal plant [12], as it exhibits a number of biological activities, such as antioxidant, anti-inflammatory [13], and antifungal [14], owing to its bioactive compounds such as terpenoids and phenylpropanoids. CJ, commonly called the Japanese cedar, is known to have strong antimicrobial [15], anti-inflammatory [16], and antifungal [17] activities. It also has skin-whitening and antioxidant activities based on anti-tyrosinase, radical scavenging, and superoxide dismutase assay results [18]. We hypothesized that a mixture of the leaf extracts of these three trees could synergistically exert skin-beneficial effects. The skin-related bioactivities of tree extracts are likely to be associated with polyphenols. Because their extraction yields and compositions can vary greatly depending on a number of factors, including solvent type and concentration, temperature, time, and extraction method, the extraction conditions need to be carefully optimized, as reported by Tanase et al. [19].

The versatile green solvents, deep eutectic solvents (DESs), have recently gained much attention due to their low toxicity, high extractability, and good biodegradability. Their potential usage in various fields, such as the cosmetics industry, has been suggested [20,21]. Therefore, the current study aims at applying DESs as safe and efficient extraction mediums that could yield maximized skin-related bioactivities from a mixture of the three long-lived trees. We therefore synthesized a number of DESs using components that are permissible according to the European Commission's Inventory of Cosmetic Ingredients (2006) [22], and compared their extraction efficiency for a selected marker compound. Then, we applied a simplex-centroid mixture design to locate the optimal combinations of CC, CJ, and GB that would maximally allow beneficial skin-related bioactivities, including antioxidant, anti-tyrosinase, and anti-elastase activities. This study suggests the potential of a mixture of the three extracts in DESs as a new natural bioactive compound-containing material for cosmeceutical product development.

2. Materials and Methods

2.1. Chemicals and Reagents

Fresh leaves of *Cinnamomum camphora* (L.) J. Presl., *Cryptomeria japonica* (L.f.) D. Don, and *Ginkgo biloba* L. were collected in Hanlim-eup, Aewol-eup, and Hawon-dong, respectively, all of which are local villages in Jeju Island, Republic of Korea. The samples were collected in October, 2018 and identified botanically by Dr. Jun Hwan Shin (Dongyang University, Yeongju, Korea). Three voucher specimens (CC: AP-0100, CJ: AP-0101, and CB: AP-0103) were deposited at the Plant Archive of Amorepacific Research and Development Center. Dried leaves of CC, CJ, and GB trees were ground into fine powders using an electric mill and stored in sealed containers at 4 °C until use. The pulverized samples were directly used for extraction without sieving. The compounds used for DES synthesis included glycerol (≥ 99.5%), xylitol (≥ 99%), D-(+)-glucose (≥ 99.5%), D-sorbitol (≥ 99.5%), D-(+)-maltose (≥ 99%), maltitol (≥ 98%), D-(-)-fructose (≥ 99%), sucrose (≥ 99.5%), betaine (≥ 98%), DL-malic acid (≥ 99%), and DL-lactic acid (~90%) (Sigma-Aldrich, St. Louis, MO, USA). They were used without further purification. The following were used for the activity assays: 2,2-di(4-tert-octylphenyl)-1-picrylhydrazyl (DPPH) and 6-hydroxy-2,5,7,8-tetramethyl chroman-2-carboxylic acid (Trolox; 97%) for the antioxidant activity; tyrosinase from mushroom, levodopa (L-DOPA), sodium phosphate dibasic, sodium phosphate monobasic, and kojic acid (KA, ≥ 99%) for the anti-tyrosinase activity; and elastase from porcine pancreas, N-succinyl-Ala-Ala-Ala-p-nitroanilide (SANA), tris base, and hydrochloric acid for the anti-elastase activity (all; Sigma-Aldrich). Isoquercetin (ISO, > 98%) (Biopurify Phytochemicals Ltd., Chengdu, China) and HPLC-grade methanol, water, and acetonitrile were also purchased (Honeywell Burdick & Jackson, Ulsan, Korea).

2.2. Sample Preparation

Tree powders (60 mg) were each extracted in 1 mL of extraction solvent with ultrasonic irradiation for 60 min. For this, an ultrasonic bath (Powersonic 410) from Hwashin Technology (Seoul, Korea) was used without temperature control. After the extraction, the bath temperature increased by approximately 20 °C. After centrifugation at 12 300 × g for 10 min, the supernatants were collected for further analyses. For ISO content determination, each supernatant was diluted (5-fold) with 70% methanol, and filtered through a 0.45 μm pore membrane syringe filter before injection into LC-UV. For the antioxidant and anti-elastase activity assays, each supernatant was diluted (10-fold) in 70% methanol, while for the anti-tyrosinase activity assay, a 10-fold diluted extract was prepared with 0.1 M phosphate buffer (pH 6.8).

2.3. Synthesis of DESs

The DESs were synthesized using the previously reported heating method [23,24], with slight modifications. In brief, accurately weighed individual components were mixed with/without a minimum amount of water and stirred at 80 °C. Then, the mixtures were kept at −80 °C for 2 h, followed by freeze drying until reaching a constant weight. All 15 DESs in Table 1 could be synthesized as homogeneous transparent liquids.

Table 1. List of synthesized deep eutectic solvents (DESs).

DES No.	Component 1	Component 2	Component 3	Molar Ratio
1 [a]		Xylitol		2:1
2 [a]		Maltose		3:1
3 [a]		Sorbitol		2:1
4 [a]		Fructose		3:1
5 [b]	Glycerol	Sucrose		3:1
6 [b]		Glucose		3:1
7 [b]		Maltitol		3:1
8 [b]		Malic acid		1:1
9 [b]		Malic acid		1:2
10 [b]	Lactic acid	Glucose		1:2
11 [b]	Fructose	Sucrose		1:1
12 [b]		Sucrose	Glucose	1:1:1
13 [b]		Sucrose		1:1
14 [b]	Betaine	Sucrose		1:2
15 [b]		Glucose		1:1

[a] Synthesized without water. [b] Synthesized using a minimum amount of water before lyophilization.

2.4. Determination of ISO as a Common Marker Compound for Extraction Efficiency Using LC-UV

Chemical profiles of the leaf extracts of the three trees were obtained using liquid chromatography-diode array detection (LC-DAD) (Figure S1), which indicated that they had different constituents and compositions. The UV spectra of the peaks in the chromatograms showed they contained a number of flavonoids, a class of components widely existing in plants, with various beneficial bioactivities (e.g., characteristic λ_{max} values at 254–264 nm and 347–353 nm). In particular, the flavonoid isoquercetin (ISO) could be identified in the extracts of all three tree leaves (Figure S1). Owing to its many beneficial bioactivities, including antioxidant [25], antivirus [26], and anti-inflammation [27] activities, ISO could serve as a marker for assessing the extraction efficiency of various extraction solvents.

A stock solution of ISO (1000 µg/mL) was prepared in methanol, and used to prepare a series of standard working solutions with concentrations ranging from 1–100 µg/mL in 70% methanol. The solutions were stored at 4 °C until use.

A Waters HPLC system (Waters Co., Milford, MA, USA) equipped with a Waters 2695 separation module and a Waters 996 photodiode array detector was used to determinate ISO levels. The injection volume was 10 µL, and chromatographic separation was achieved on a Gemini C_{18} column (250 × 4.6 mm, 5 µm; Phenomenex, Torrance, CA, USA) at room temperature. The detector wavelength was set at 354 nm. The mobile phase consisted of eluent A (1% formic acid in water) and eluent B (1% formic acid in acetonitrile) run at a flow rate of 1 mL/min. Gradient elution was performed as follows: 0–3 min, 5–20% of B; 3–30 min, 15–20% of B; 30–31 min, 20–100% of B; 31–37 min, 100% of B; and B % was returned to 5% in 1 min. The established calibration equation was y = 20076x + 1013 (r^2 = 1.0000, n = 3).

2.5. Antioxidant Activity Assay

DPPH free radical scavenging activity was assessed using the previously established protocol [23]. Briefly, the supernatant was diluted (10-fold) in 70% methanol, and mixed with an equal volume of freshly prepared 0.2 mM DPPH in methanol. Absorbance at 517 nm was measured after a 30 min incubation of the mixture seeded in a 96-well microplate in the dark at room temperature. The antioxidant abilities of the extracts were calculated using the following equation:

$$\text{Scavenging activity\%} = \left(A_{control} - A_{sample}\right)/A_{control} \times 100 \tag{1}$$

The radical scavenging activity was expressed as Trolox equivalents (mg TE/g tree leave sample). The linear calibration range for Trolox was 1–20 µg/mL, with y = 4.160 x + 2.955 (r^2 = 0.9982, n = 3).

2.6. Anti-Tyrosinase Activity Assay

The anti-tyrosinase activities of diluted samples (10-fold) were measured with L-DOPA as the substrate, as previously reported with slight modifications [28]. A mixture of 100 µL of the test samples and 100 µL of 80.61 U/mL of tyrosinase in 0.1 M phosphate buffer (pH 6.8) was incubated at room temperature for 15 min. Then, 50 µL of 2.5 mM L-DOPA was added and the mixture was incubated at room temperature for 20 min. Absorbance was measured at 475 nm. Enzyme inhibitory activity was calculated using the following equation:

$$\text{Inhibition\%} = (1 - (A_1 - A_2)/(A_3 - A_4)) \times 100 \tag{2}$$

where, A_1, A_2, A_3, and A_4 are the absorbances of test sample with enzyme, test sample without enzyme, enzyme without test sample, and solution without test sample and enzyme, respectively.

2.7. Anti-Elastase Activity Assay

SANA was used as a substrate in the anti-elastase activity assay, conducted as previously reported with minor modifications [29]. First, 50 µL of 10-fold diluted test sample was mixed with 100 µL of 10 µg/mL elastase in 0.2 M Tris-HCl buffer (pH 8.0), and incubated at room temperature for 15 min. Then, 100 µL of substrate (1.6 mM SANA dissolved in 0.2 M Tris-HCl buffer, pH 8.0) was added to the mixture and incubated for 20 min. Absorbance was measured at 405 nm, and Equation (2) was used to calculate % inhibition. All activity assays were performed in triplicate.

2.8. Experimental Design and Statistical Analyses

Design Expert 8.0 software (Statease Inc., Minneapolis, MN, USA) was used for the simplex-centroid mixture design, and for analyzing data and creating graphs. All experiments were performed in random order. Other statistical analyses, including multiple comparisons, were conducted using GraphPad Prism 6 for Windows (GraphPad Software, San Diego, CA, USA).

3. Results and Discussion

3.1. Preparation of DESs

A list of components was selected from the European Commission's Inventory of Cosmetic Ingredients (2006), considering their stability, price, and safety, and then combined at proper molar ratios with hydrogen bond acceptors (HBAs) and hydrogen bond donors (HBDs), based on the available literature [20,23,24]. As a result, a total of 15 DESs were successfully synthesized using the heating method (Table 1).

3.2. Selection of DESs for Mixture Design Studies Based on the ISO Extraction Yields

The ISO content of CC, CJ, and GB were determined to be 861, 276, and 118 µg/g, respectively, when using 70% w/w DES 1. Because CC had the highest ISO level among the three trees, it was therefore used as a model for comparing the extraction efficiency of the prepared DESs. The ISO levels post extraction with DESs 1–11 (746–915 µg/g) were significantly ($p < 0.05$) higher than those with hot water (705 µg/g) (Figure 1). However, they were significantly ($p < 0.05$) lower than those acquired using 70% methanol (1086 µg/g). Glycerol generally served as a good component for DESs that are effective for ISO extraction, whether its counterpart was a sugar or an organic acid. Moreover, organic acids appeared to be superior components to glycerol, and DESs 8–10 containing malic acid or lactic acid resulted in the highest ISO extraction yields. However, DESs with organic acids are usually highly viscous and thus difficult to handle [24], and so were the malic acid-based DESs (DESs 8 and 9). DES 10, with yield similar to those of DESs 8 and 9, was excluded from further consideration because of the low lactic acid purity (~90%). Accordingly, DES 1 (ISO yield, 861 µg/g), which could be synthesized from glycerol and xylitol and handled with ease, was deemed the reasonable extraction solvent for the tree leaves.

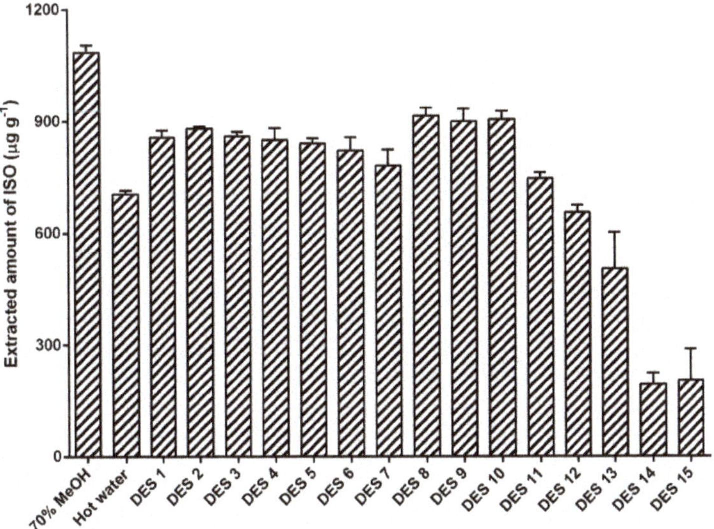

Figure 1. Comparison of isoquercetin (ISO) levels post extraction with DESs, to those post extraction with 70% methanol and water. The water content of all the tested DESs was 30% w/w. Error bars indicate standard error of the mean (n = 3).

Before applying DES 1 to the mixture design studies, the effects of several factors on extraction efficacy were examined. First, in varying the molar ratio of glycerol to xylitol from 1:1 to 5:1, at 30% water, we found that DES 1 with a 5:1 ratio (designated as DES 1-5) had the highest efficiency, although it was not significantly different from that of DES 1 with a 4:1 ratio (DES 1-4) (Table S1). Then, the effect of the water content of DES 1 was assessed by measuring ISO yields with DES 1 with various water contents (10–50% w/w). We found that 30% water exhibited the highest yield (946 µg/g), while 50 and 10% water resulted in <900 µg/g and ~700 µg/g, respectively (Table S2). Based on the faster synthesis time and lower viscosity, DES 1-5 was regarded as desirable, and its 70% aqueous solution was used as the final extraction solvent in the mixture design studies. It is apparent that the selected solvent (70% w/w DES 1-5) is less economical than water as an extraction medium. However, it is noteworthy that the extraction efficiency of the selected solvent was achieved without any heat treatment, and that the individual DES components can act as functional ingredients in cosmetics [20]. The water extraction, which had to be conducted at a boiling temperature for improved efficiency, might cause degradation of unstable compounds during extraction. These factors could provide merit for the use of the DES solution in this study.

3.3. Simplex-Centroid Mixture Design

Using the generated design matrix, nine different mixtures were prepared with CC, CJ, and GB, extracted in 70% DES 1-5, and their antioxidant (Y_1), anti-tyrosinase (Y_2), and anti-elastase (Y_3) activities were evaluated. The design matrix with coded and real values and the resulting responses is summarized in Table 2. For better evaluation of the design, the central point was repeated thrice [30,31].

Table 2. Simplex-centroid mixture design and the resulting responses.

No.	Independent Variables [a]			Dependent Variables [b]		
	X_1 (mg)	X_2 (mg)	X_3 (mg)	Y_1	Y_2	Y_3
1	1 (60)	0 (0)	0 (0)	3.259	60.44	72.11
2	0 (0)	1 (60)	0 (0)	3.175	68.35	78.86
3	0 (0)	0 (0)	1 (60)	3.175	39.61	50.67
4	0.5 (30)	0.5 (30)	0 (0)	3.246	50.55	76.96
5	0.5 (30)	0 (0)	0.5 (30)	3.252	62.48	82.92
6	0 (0)	0.5 (30)	0.5 (30)	3.131	30.50	9.83
7	0.33 (20)	0.33 (20)	0.33 (20)	3.271	54.99	62.64
8	0.33 (20)	0.33 (20)	0.33 (20)	3.252	50.98	72.07
9	0.33 (20)	0.33 (20)	0.33 (20)	3.258	42.50	73.04

[a] X_1, *Cinnamomum camphora* (CC); X_2, *Cryptomeria japonica* (CJ); X_3, *Ginkgo biloba* (GB). [b] Y_1, antioxidant activity (mg TE/g); Y_2, anti-tyrosinase activity (% inhibition); Y_3: anti-elastase activity (% inhibition).

The fitness of the linear, quadratic, and special cubic models for the three responses were analyzed (Table S3). The models with highest R^2_{adj} values with statistical significance ($p < 0.05$) were selected for each response, resulting in the special cubic, quadratic, and special cubic models for Y_1, Y_2^3, and Y_3, respectively. The model equations are displayed in Table 3, and the model qualities evaluated using analysis of variance (ANOVA) are summarized in Table S4.

Table 3. Model equations describing response as a function of mixture composition.

Response [a]	Model	R^2	R^2_{adj}
Y_1	$3.26X_1 + 3.17X_2 + 3.17X_3 + 0.12X_1X_2 + 0.14X_1X_3 - 0.18X_2X_3 + 1.31X_1X_2X_3$	0.9905	0.9619
Y_2^3 [b]	$219231.94X_1 + 317719.32X_2 + 60577.95X_3 - 531992.97X_1X_2 + 441173.67X_1X_3 - 617876.6X_2X_3$	0.9400	0.8399
Y_3	$8.49X_1 + 8.88X_2 + 7.12X_3 + 0.35X_1X_2 + 5.2X_1X_3 - 19.46X_2X_3 + 45.87X_1X_2X_3$	0.9911	0.9645

[a] Y_1, antioxidant activity; Y_2, anti-tyrosinase activity; Y_3, anti-elastase activity. [b] Transformation: Power (Lambda = 3; constant, k = 0).

For the antioxidant activity (Y_1), only a special cubic model was significant (95% confidence level). The linear terms and cubic term had significant positive effects on the antioxidant activity (Table S4). As displayed in the contour plot (Figure 2a) and the model equation (Table 3), the X_1 variable (CC) provided a greater contribution than X_2 (CJ) and X_3 (GB). Specifically, higher responses were observed as the X_1 portion of the mixture increased. The model yielded the following optimized proportions: X_1, 53.3%; X_2, 23.3%; and X_3, 23.3%.

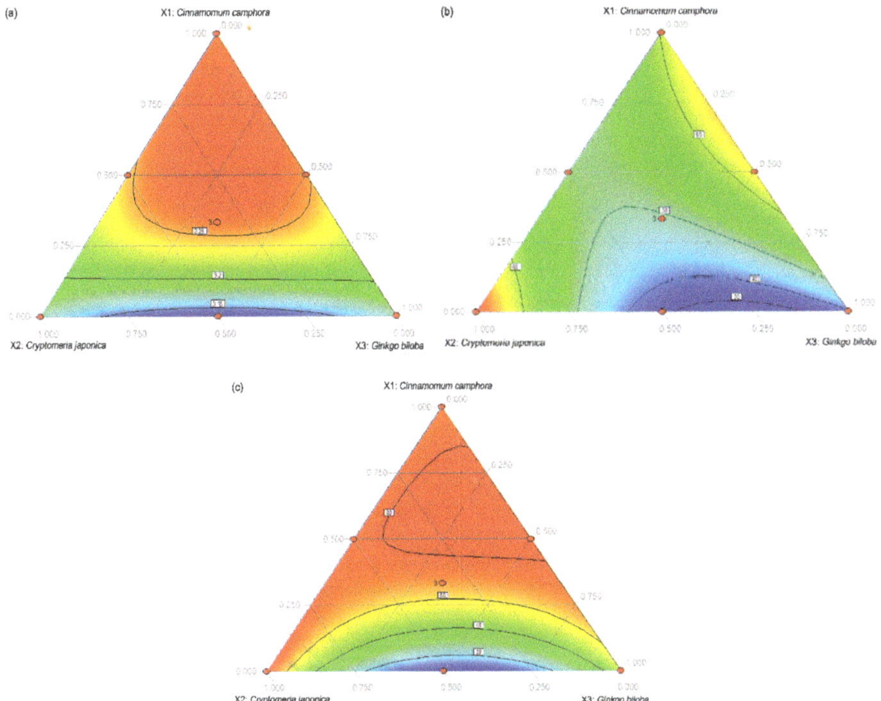

Figure 2. Response surface contour plots for the (**a**) antioxidant, (**b**) anti-tyrosinase, and (**c**) anti-elastase activities of different compositions of X_1 (CC), X_2 (CJ), and X_3 (GB).

For the anti-tyrosinase activity (Y_2), response transformation was vital to obtaining a significant model with no significant lack of fit, using the transformation type "power" with lambda and constant k set as +3 and 0, respectively. The resulting quadratic model was significant, in which only the two quadratic terms X_1X_2 and X_2X_3 were significant, as the term $X_1X_2X_3$ was excluded from modeling due to its negligible contribution (Table S4). X_2 (CJ) generally enhances the anti-tyrosinase activity, while X_1 and X_2, and X_2 and X_3 inhibit the anti-tyrosinase activity (Figure 2b and Table 3). The model suggested the use of 100% X_2 to achieve the highest anti-tyrosinase activity.

For the anti-elastase activity, only a special cubic model was significant, of which the linear terms, the quadratic term X_2X_3, and the cubic term were significant (Table S4). X_1 (CC) had the highest anti-elastase activity-enhancement effect, while the mixture of X_2 and X_3 inhibited anti-elastase activity (Figure 2c and Table 3). According to the model, the optimal compositions for achieving the highest anti-elastase activity were 63.6, 14.1, and 22.3% for X_1, X_2, and X_3, respectively.

4. Conclusions

The skin-beneficial bioactivities of the mixtures of extracts from three kinds of long-lived trees were investigated to identify potential new cosmeceutical product materials. Extracts of their leaves extracted using DESs contained cosmetics-compatible compounds, allowing safe and efficient ISO extraction into the extracts and their direct application in cosmetic formulations. Various mixtures of the three leaf extracts were prepared and systematically analyzed according to the simplex-centroid mixture design, providing information about the effects of the individual leaves on the three different bioactivities, as well as the optimal compositions of the three leaves for maximized antioxidant, anti-tyrosinase, and anti-elastase activities. The strategy used in this study could be applied to exploring new, effective, and safe materials for the cosmetics and related fields. However, further investigation of potential toxicity profiles for the selected DES would be needed before application, because DESs could have different toxicity than their individual components that have no or low toxicity.

Supplementary Materials: The following are available online at http://www.mdpi.com/2076-3417/9/13/2581/s1, Table S1. Effects of the molar ratios between glycerol and xylitol on the ISO yields; Table S2. Effects of the added water in DES 1-5 (glycerol:xylitol, 5:1); Table S3. Fit summary of the models; Table S4. ANOVA results of the established models for each response; Figure S1. Chromatographic profiles of the three extracts of (a) *Ginkgo biloba*, (b) *Cinnamomum camphora*, and (c) *Cryptomeria japonica* leaves prepared using 70% w/w DES 1. Peak identification: ISO, isoquercetin.

Author Contributions: Y.J., M.Y., J.K., and J.L. conceived and designed the experiments; Y.J., D.J., K.L., M.Y., J.K., and K.P. performed the experiments. Y.J. analyzed the data and drafted the manuscript; J.L. revised and edited the manuscript.

Funding: This work was financially supported by a grant from the Amorepacific Corporation (grant No.ORT-01-R18E999005) within the Amorepacific Open Research Program.

Conflicts of Interest: The authors declare no conflict of interest. J.K. and M.Y. work for Amorepacific Corporation.

References

1. Masaki, H. Role of antioxidants in the skin: Anti-aging effects. *J. Dermatol. Sci.* **2010**, *58*, 85–90. [CrossRef] [PubMed]

2. McDaniel, D.; Farris, P.; Valacchi, G. Atmospheric skin aging—Contributors and inhibitors. *J. Cosmet. Dermatol.* **2018**, *17*, 124–137. [CrossRef] [PubMed]

3. Shivanand, P.; Nilam, M.; Viral, D. Herbs play an important role in the field of cosmetics. *Int. J. PharmTech Res.* **2010**, *2*, 632–639.

4. Juhász, M.L.; Levin, M.K.; Marmur, E.S. The use of natural ingredients in innovative Korean cosmeceuticals. *J. Cosmet. Dermatol.* **2018**, *17*, 305–312. [CrossRef]

5. Eyles, A.; Bonello, P.; Ganley, R.; Mohammed, C. Induced resistance to pests and pathogens in trees. *New Phytol.* **2010**, *185*, 893–908. [CrossRef] [PubMed]

6. Ding, S.; Dudley, E.; Plummer, S.; Tang, J.; Newton, R.P.; Brenton, A.G. Quantitative determination of major active components in Ginkgo biloba dietary supplements by liquid chromatography/mass spectrometry. *Rapid Commun. Mass Spectrom.* **2006**, *20*, 2753–2760. [CrossRef] [PubMed]

7. Shu, Z.; Shar, A.H.; Shahen, M.; Wang, H.; Alagawany, M.; El-Hack, M.E.A.; Kalhoro, S.A.; Rashid, M.; Shar, P.A. Pharmacological Uses of Ginkgo Biloba Extracts for Cardiovascular Disease and Coronary Heart Diseases. *Int. J. Pharmacol.* **2019**, *15*, 1–9.

8. van Beek, T.A.; Montoro, P. Chemical analysis and quality control of Ginkgo biloba leaves, extracts, and phytopharmaceuticals. *J. Chromatogr. A* **2009**, *1216*, 2002–2032. [CrossRef] [PubMed]

9. van Beek, T.A. Chemical analysis of Ginkgo biloba leaves and extracts. *J. Chromatogr. A* **2002**, *967*, 21–55. [CrossRef]

10. Zuo, W.; Yan, F.; Zhang, B.; Li, J.; Mei, D. Advances in the studies of Ginkgo biloba leaves extract on aging-related diseases. *Aging Dis.* **2017**, *8*, 812. [CrossRef]

11. Chan, P.-C.; Xia, Q.; Fu, P.P. *Ginkgo biloba* leave extract: Biological, medicinal, and toxicological effects. *J. Environ. Sci. Health C* **2007**, *25*, 211–244. [CrossRef] [PubMed]

12. Chen, W.; Vermaak, I.; Viljoen, A. Camphor—A fumigant during the black death and a coveted fragrant wood in ancient Egypt and Babylon—A review. *Molecules* **2013**, *18*, 5434–5454. [CrossRef] [PubMed]

13. Li, Y.-R.; Fu, C.-S.; Yang, W.-J.; Wang, X.-L.; Feng, D.; Wang, X.-N.; Ren, D.-M.; Lou, H.-X.; Shen, T. Investigation of constituents from *Cinnamomum camphora* (L.) J. Presl and evaluation of their anti-inflammatory properties in lipopolysaccharide-stimulated RAW 264.7 macrophages. *J. Ethnopharmacol.* **2018**, *221*, 37–47. [CrossRef]

14. Chen, Y.; Dai, G. Antifungal activity of plant extracts against Colletotrichum lagenarium, the causal agent of anthracnose in cucumber. *J. Sci. Food Agric.* **2012**, *92*, 1937–1943. [CrossRef]

15. Cha, J.D.; Jeong, M.R.; Jeong, S.I.; Moon, S.E.; Kil, B.S.; Yun, S.I.; Lee, K.Y.; Song, Y.H. Chemical composition and antimicrobial activity of the essential oil of *Cryptomeria japonica*. *Phytother. Res.* **2007**, *21*, 295–299. [CrossRef]

16. Shyur, L.-F.; Huang, C.-C.; Lo, C.-P.; Chiu, C.-Y.; Chen, Y.-P.; Wang, S.-Y.; Chang, S.-T. Hepatoprotective phytocompounds from *Cryptomeria japonica* are potent modulators of inflammatory mediators. *Phytochemistry* **2008**, *69*, 1348–1358. [CrossRef] [PubMed]

17. Cheng, S.-S.; Lin, H.-Y.; Chang, S.-T. Chemical composition and antifungal activity of essential oils from different tissues of Japanese cedar (*Cryptomeria japonica*). *J. Agric. Food Chem.* **2005**, *53*, 614–619. [CrossRef]

18. Kim, S.; Lee, S.; Hong, C.; Gwak, K.; Park, M.; Smith, D.; Choi, I. Whitening and antioxidant activities of bornyl acetate and nezukol fractionated from *Cryptomeria japonica* essential oil. *Int. J. Cosmet. Sci.* **2013**, *35*, 484–490. [CrossRef] [PubMed]

19. Tanase, C.; Domokos, E.; Cosarca, S.; Miklos, A.; Imre, S.; Domokos, J.; Dehelean, C.A. Study of the ultrasound-assisted extraction of polyphenols from beech (*Fagus sylvatica* L.) bark. *Bioresources* **2018**, *13*, 2247–2267. [CrossRef]

20. Jeong, K.M.; Ko, J.; Zhao, J.; Jin, Y.; Yoo, D.E.; Han, S.Y.; Lee, J. Multi-functioning deep eutectic solvents as extraction and storage media for bioactive natural products that are readily applicable to cosmetic products. *J. Clean Prod.* **2017**, *151*, 87–95. [CrossRef]

21. Dai, Y.; Verpoorte, R.; Choi, Y.H. Natural deep eutectic solvents providing enhanced stability of natural colorants from safflower (*Carthamus tinctorius*). *Food Chem.* **2014**, *159*, 116–121. [CrossRef] [PubMed]

22. The Commission of the European Communities. Commission Decision of 9 February 2006 amending Decision 96/335/EC establishing an inventory and a common nomenclature of ingredients employed in cosmetic products. *Off. J. Eur. Union* **2006**, *257*, 1–528.

23. Yoo, D.E.; Jeong, K.M.; Han, S.Y.; Kim, E.M.; Jin, Y.; Lee, J. Deep eutectic solvent-based valorization of spent coffee grounds. *Food Chem.* **2018**, *255*, 357–364. [CrossRef]

24. Dai, Y.; van Spronsen, J.; Witkamp, G.-J.; Verpoorte, R.; Choi, Y.H. Natural deep eutectic solvents as new potential media for green technology. *Anal. Chim. Acta* **2013**, *766*, 61–68. [CrossRef] [PubMed]

25. Li, R.; Yuan, C.; Dong, C.; Shuang, S.; Choi, M.M. In vivo antioxidative effect of isoquercitrin on cadmium-induced oxidative damage to mouse liver and kidney. *Naunyn Schmiedebergs Arch. Pharmacol.* **2011**, *383*, 437–445. [CrossRef]

26. Kim, Y.; Narayanan, S.; Chang, K.-O. Inhibition of influenza virus replication by plant-derived isoquercetin. *Antivir. Res.* **2010**, *88*, 227–235. [CrossRef]

27. Ma, C.; Jiang, Y.; Zhang, X.; Chen, X.; Liu, Z.; Tian, X. Isoquercetin ameliorates myocardial infarction through anti-inflammation and anti-apoptosis factor and regulating TLR4-NF-κB signal pathway. *Mol. Med. Rep.* **2018**, *17*, 6675–6680. [CrossRef]

28. Chatatikun, M.; Chiabchalard, A. Thai plants with high antioxidant levels, free radical scavenging activity, anti-tyrosinase and anti-collagenase activity. *BMC Complement. Altern. Med.* **2017**, *17*, 487. [CrossRef] [PubMed]

29. Sundaram, I.K.; Sarangi, D.D.; Sundararajan, V.; George, S.; Mohideen, S.S. Poly herbal formulation with anti-elastase and anti-oxidant properties for skin anti-aging. *BMC Complement. Altern. Med.* **2018**, *18*, 33.

30. Handa, C.L.; de Lima, F.S.; Guelfi, M.F.G.; Georgetti, S.R.; Ida, E.I. Multi-response optimisation of the extraction solvent system for phenolics and antioxidant activities from fermented soy flour using a simplex-centroid design. *Food Chem.* **2016**, *197*, 175–184. [CrossRef]

31. Bau, T.R.; Garcia, S.; Ida, E.I. Optimization of a fermented soy product formulation with a kefir culture and fiber using a simplex-centroid mixture design. *Int. J. Food Sci. Nutr.* **2013**, *64*, 929–935. [CrossRef] [PubMed]

 *applied sciences*

Review

Deep Eutectic Solvents as Extraction Media for Valuable Flavonoids from Natural Sources

Dimitris Skarpalezos and Anastasia Detsi *

Laboratory of Organic Chemistry, Department of Chemical Sciences, School of Chemical Engineering, National Technical University of Athens, 15780 Zografou, Athens, Greece; dskarpalezos@hotmail.com
* Correspondence: adetsi@chemeng.ntua.gr

Received: 31 July 2019; Accepted: 30 September 2019; Published: 4 October 2019

Abstract: The present review article attempts to summarize the use of deep eutectic solvents in the extraction of flavonoids, one of the most important classes of plant secondary metabolites. All of the applications reviewed have reported success in isolation and extraction of the target compounds; competitive, if not superior, extraction rates compared with conventional solvents; and satisfactory behavior of the extract in the latter applications (such as direct analysis, synthesis, or catalysis), wherever attempted.

Keywords: flavonoids; extraction; deep eutectic solvents (DES); natural deep eutectic solvents (NaDES)

---

1. Introduction

The global turn towards green chemistry is an integral approach towards conventional chemical practices. Following this trend, both raw materials and processes are being re-evaluated from the ground up, combining the research for naturally, sustainably sourced raw materials with eco-friendly, cost effective, and lean processes.

Research on multiple compound groups is being carried out globally, aiming at isolating substances with significant health, well-being, or other benefits from natural sources, through green methods.

The present review focuses on the extraction of flavonoids, an immense category of compounds found in plants and natural products. Furthermore, the review is dedicated to conventional and novel extraction methods using classic or natural deep eutectic solvents, a new category of green solvents with exceptional solvent properties, as well as generally green behavior.

An inclusive overview of the current research being carried out on the extraction of flavonoids using deep eutectic solvents is provided. The main conclusions, issues, and trends are analyzed in order to comprise a solid foundation for further research or more focused application.

1.1. Flavonoids

Flavonoids are a category of naturally occurring organic compounds found in fruits, vegetables, or grains. Flavonoids, as a group, include upwards of 8000 different identified compounds, which are responsible for a number of functions within plants, such as the coloration of the different parts of fruits or vegetables (leaves, flower, peel), as well as shielding against UV rays or external threats, such as pathogens. They can be located in other subsystems as well, such as the bark or the roots of the plant, where they might serve similar or different purposes [1].

The interest around extraction of flavonoids stems from the multiple health benefits they provide. Cardiovascular benefits, anticancer activity, and neurological system fortification are only some of the many actions this group. Flavonoid intake has been involuntarily pursued in medicine since ancient times. Herbal medicine, such as traditional Chinese or Mediterranean medicine, has helped breed plants that are considerably rich in flavonoids. This legacy has fueled much of the research listed in this project, directing green extraction towards traditional herbs, fruits, or common foods in the hopes of isolating useful substances, including flavonoids [2,3].

From a chemical structure standpoint, all flavonoids stem from a main skeleton that they share, and differentiate from each other based on the substituents attached to any part of the structure (Figure 1). They possess phenolic and pyrane rings in their structures and have many subclasses, such as flavanols, flavones, flavanones, chalcones, and anthocyanidines. The flavonoid skeleton is comprised of an aromatic ring linked on one side with a six-membered heterocyclic ring, which bears an oxygen atom instead of carbon next to the common side. The two rings connect to another aromatic ring to form the skeleton, as shown below:

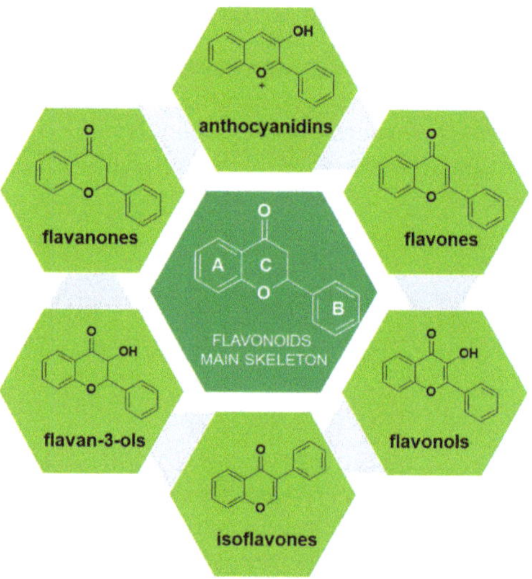

Figure 1. Flavonoid skeleton and general structures of the main flavonoid categories.

1.2. Deep Eutectic Solvents

The extraction media of choice for green extractions could very well be Deep Eutectic Solvents (DES). DES are solvents that occur when a mixture of substances has a melting point that is much lower than that of the two constituents. In order to form a DES system, there needs to be a hydrogen bond donor (HBD) and a hydrogen bond acceptor (HBA), which when mixed at proper ratios create a new "mesh" of hydrogen-bond-interconnected molecules with interesting physicochemical properties. DES can be highly viscous, inhibiting their use in processes that require diffusion or flow, however, research into their chemical structure, as well as the use of additives (like water), largely alleviates this issue [4].

The potential to use multiple molecules as constituents and create a fluid mixture that can be used as a solvent is the first key interesting point regarding DES. Adding to this notion, the potential for many naturally occurring molecules to form DES, thus providing a natural solvent system with low vapor pressure; low cost, even at larger industrial scales; and the potential to remove the need for solvent retrieval (Natural DES–NaDES could remain in a consumable end product) highlights their importance. The evolution of DES, and more importantly, NaDES, means that the design of the newer solvents can focus on a solvent that is a capable and biocompatible storage media, a readily available active ingredient, a very efficient catalyst, or a molecular or compound carrier [5,6].

Competing with ionic liquids (ILs), another category of green solvents so far favors DES, since they are largely cheaper to produce, less toxic, and offer great variety. Seemingly, NaDES is the natural step forward from conventional ionic liquids, since they can be formed from green sources, can be polar or non-polar, hydrophilic or hydrophobic, and can be tailored for any situation. DES could become designer solvents, designed and synthesized for every extraction individually, with maximum efficiency in mind.

DES can be synthesized via a number of methods. Considering the research listed in this review, predominant methods are often simple, such as stirring at room to high temperatures (80 °C), freeze-dry mixing, or ultrasound-assisted mixtures [7,8]. The synthesis method can be selected with cost in mind (lower temperatures prevail), speed or efficiency (favoring higher temperatures and/or ultrasound assistance) or limited by the properties of the reagents (thermal-sensitive substances might require freeze-drying instead of heating due to thermal instability).

Compounds that are preferred for their natural origin and green character, and that have been heavily employed in NaDES synthesis are shown in Table 1. All the molecules in Table 1 have been a part of DES used in flavonoid extraction.

Table 1. Common ingredients for deep eutectic solvents (DES) used in flavonoid extraction (based upon the literature cited in this review).

Hydrogen Bond Donors (HBD)	Hydrogen Bond Acceptors (HBA)
Urea	Choline Chloride
Glycerol	Betaine Chloride
Ethylene Glycol	L-Proline
1,2-Butanediol	Sodium Propionate
1,3-Butanediol	
1,4-Butanediol	
2,3-Butanediol	
1,6-Hexanediol	
Malic Acid	
Malonic Acid	

Table 1. *Cont.*

Hydrogen Bond Donors (HBD)	Hydrogen Bond Acceptors (HBA)
Citric acid 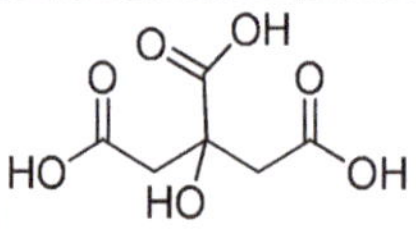	
Levulinic Acid 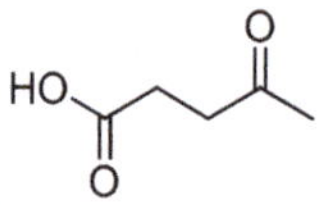	
Fructose 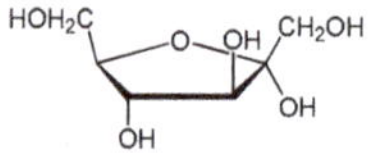	
Glucose 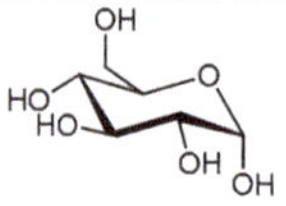	
Sucrose 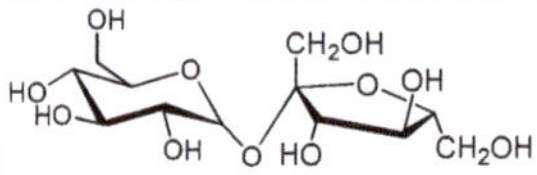	
Sorbitol 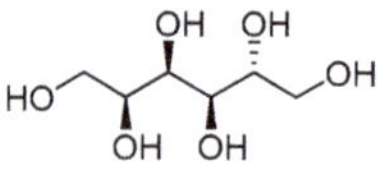	

1.3. Application of DES on Flavonoid Extractions

This review examined multiple research topics using DES in various methods to extract flavonoids from natural sources. All of the extraction methods agree upon the potential of DES in the extraction of plant sources, given the success of the extractions as well as the promising results in relation to conventional solvents, particularly so wherever a comparison could be directly drawn [9,10].

The extractions might not have the collection of a compound as the end goal. Many aimed at providing better analytical methods through the use of DES [11–14], while others attempted the creation of a substrate for several reactions or subsequent extractions (through systems constructed with the aid of DES) [15–17].

Furthermore, all of the projects have been successful at applying DES to achieve the target extractability or otherwise consequent properties, boasting either high retrievability (for extraction), sufficiently low detection limits for analytical methods, or effective substrates and precursors for reactions. Whether utilizing DES in novel methods, attempting to increase efficiency with new approaches, such as negative pressure cavitation (NPC) [18] and these novel solvents, or applying them to older extraction systems, DES generally seem to adapt to the needs of each process.

All of the applications that incorporated the extraction of a substance through DES, at some point and to some degree, have provided useful information towards creating a clear picture as to what affects the process of extraction through DES, and more importantly, how this happens.

2. Factors Affecting the Extraction of Flavonoids Using DES Separation Techniques

The careful study of the current literature concerning the extraction of flavonoids using DES reveals a set of factors, namely, temperature, molecular structure and composition of the DES, extraction time, water content, the use of additives, solvent/sample ratio, and pH, which plays an important role in the efficiency and yield of the process.

2.1. Temperature

The temperature in which the extraction takes place is naturally expected to affect the time of the extraction, as well as its efficiency and performance. In general, higher temperatures increase molecular mobility and allow extracted molecules to diffuse to the solvent quicker. Extractions through DES are no exception. More so, extractions with DES rely on temperature to decrease the viscosities of DES, which are significantly high and render extractions cumbersome.

According to the literature, the desirable temperature range for extractions is from room temperature (25 °C) to about 60 °C. Higher temperatures, aside from demanding energy to sustain (moving away from the green character of extractions), might also endanger either the DES or the target substance, since many of the natural substances involved are thermally sensitive [19].

Very high temperatures also proved to decrease yield in some cases because of a decrease in the interactability between the target compound and the solvent of choice, regardless of the thermal endurance of either one (which is still a limiting factor nevertheless) [20].

2.2. Molecular Structure and Composition of the DES

The molecular structure of the DES refers to the ingredients used in its synthesis. Whether a binary or a ternary system, the molecules contained in the DES are responsible for its unique properties. In extractions, it is reported that the polarity of the DES is a very important factor affecting solubility. In screening multiple DES against many samples and target compounds, the conclusion implies that the polarity of the DES used needs to be close to the polarity of the target substance. Among similar DES, one with a polarity closer to the target will present the greatest extractability. Therefore, in selecting the proper DES for extraction, polarity similarity is a top priority as far as efficiency is concerned. This might be difficult given the many potential structures of DES, but potentially enables great efficiency through novel structures that approach the polarity of each given target [21,22].

Furthermore, the molecular interactions of both HBDs and HBAs with the target, as well as the background, need to be considered. Any competitive interactions between the ingredients of the DES and another presence in the system might interfere with the extraction efficiency to a great degree, potentially leading to a redesign of the extraction [23,24]. An example of this occurrence comes from Cui et. al, where a change in the ratio between the donor and acceptor led to decrease in yield. This decrease was not due to polarity change or some similar factor, but due to the chloride anion that the choline and betaine carried, and which reacted with the target. Reducing their presence in the DES reduced the yield simply because of the decrease in interaction between the fewer chloride anions and the target compound [20]. Within a more general scope, with regards to the ratio of HBA/HBD, it seems that an increase in the hydrogen bond donor content leads to a decrease in the viscosity of DES. In addition, an increase of hydroxyl groups in any ingredient of the DES would promote the formation of hydrogen bonds yielding a significantly more stable DES [22].

2.3. Toxicity

Generally, DES are reported as "safe" and "non-toxic" or of "low toxicity", without any other justification than the safety and low toxicity of their components. However, this assumption can be true only in the case of NaDES, which are constituted by naturally occurring compounds, and therefore can be considered as inherently non-toxic. Thus, the literature concerning the toxicity of DES or NaDES is still scarce, and in the majority of the published works involving their use as extraction solvents, no toxicity tests are included.

Hayyan et al. [25] were among the first to study the toxicity and cytotoxicity of DES possessing choline chloride as the HBA and glycerine, ethylene glycol, triethylene glycol, and urea as the HBD. The tested DES and their individual components did not show toxicity against *Bacillus subtilis*, *Staphylococcus aureus*, *Escherichia coli*, or *Pseudomonas aeruginosa*. The interesting finding was that the DES under study showed significantly higher cytotoxicity than their individual components against *Artemia salina* leaches. This striking difference in cytotoxicity was attributed to the hydrogen bonding network present in DES and definitely merits further investigation.

The toxicity of a series of NaDES against L929 fibroblast-like cells was studied by Duarte et al. [26]. The results indicated that although no clear trend regarding cytotoxicity in relation with structure was observed, the presence of organic acids as HBDs results in increased cytotoxic activity. A series of 28 NaDES containing ChCl as the HBA and a variety of HBDs were tested for their cytotoxicity against the human embryonic kidney cell line (HEK-293) [27]. The results showed that ChCl, as well as the compounds used as HBDs, were less toxic than the corresponding NaDES and that the structure of the HBD and the HBA/HBD ratio play a role in cytotoxicity.

Radošević et al. [28,29] studied different NaDES regarding their antimicrobial activity against *Salmonella typhimurium.*, *Escherichia coli*, *Pseudomonas aeruginosa*, *Proteus mirabilis*, *Staphylococcus aureus*, and *Candida albicans* as well as their cytotoxicity against human normal and cancer cell lines (HEK293T, HeLa, MCF-7). NADES containing an organic acid were found to possess good antimicrobial activity, whereas their individual components were not active. Moreover, the majority of the tested NaDES showed low cytotoxicity, with the exception of ChCl-oxalic acid, which exhibited moderate cytotoxicity selectively against cancer cells. This observation is very important and the authors claim that this can be attributed to the fact that cancer cells have higher energy demands than normal cells.

In the work of Macario et al. [30], a series of DES comprised of ChCl, tetramethylammonium chloride ([N1111]Cl), and tetrabutylammonium chloride ([N4444]Cl) as HBAs, in combination various HBDs, were extensively studied for their cytotoxicity against two human skin cell lines, HaCaT32–35 (chosen as model for cosmetic applications) and MNT-136–38 (selected as a model to understand the potential of the DES under study for the treatment of skin disorders). The ChCl- and [N1111]Cl-containing DES were not cytotoxic, and some of them even increased cell viability. Thus, these DES can be safely characterized as "benign", at least for these cell lines, and for skin-related applications. The [N4444]Cl containing DES was cytotoxic against these cell lines, and as no clear trend regarding the relation of cytotoxicity with the HBD used was deduced, this HBA should not be considered safe for further applications.

The in vivo safety of DES and NaDES is much less studied. The first published research in which both in vitro and in vivo toxicity of DES was conducted is the work of Hayyan et al. in 2015 [31]. Four DES possessing ChCl as the HBA along with glycerine, ethylene glycol, triethylene glycol, and urea were tested against five human cancer cell lines and one normal cell line, and the individual components were tested as well. The cytotoxicity of DES in the various cell lines was found to be not negligible, and the ChCl/HBD ratio as well as the HBD structure seem to play important roles in toxicity. The in vitro acute toxicity studies indicated that the examined DES were more toxic than their individual components.

The cytotoxicity of two NaDES having ChCl as the HBA and glucose and fructose as HBDs, as well as the DES N,N-diethylethanol ammonium chloride-triethylene glycol, was studied by Mbous et al. [32] against 6 cancer cell lines. NaDES were found to be less toxic than the DES in the in vitro tests. In the in vivo tests, however, the NaDES showed higher toxicity than the DES, a result that was attributed to the higher viscosity of the NaDES.

Toxicological studies of DES and NaDES should be conducted before they are used in any application involving administration to living organisms, animals, or humans. In this context, Chen et al. [33] performed an acute toxicity study to test the safety of ChCl-glycerine DES, which was to be used as a drug carrier for salvianolic acid B. They were grateful to find that the LD50 value (Median Lethal Dose) of the tested DES was 7733 mg/kg, with a 95% confidence interval of 7130–8387 mg/kg for oral administration; thus, it can be safely administered orally, as it did not promote acute toxicity.

Belebna et al. [34] recently published the toxicity evaluation of an extract from green coffee beans rich in polyphenolic compounds. The extraction medium was NaDES (betaine-glycerol) and the studies were conducted in vivo on rats in order to investigate the potential of administering the extract as a dietary supplement. The NaDES extract induced several adverse effects after a high dose was administered orally, and the authors correctly highlight that in the case of developing food supplements based on NaDES extracts, the dose should be carefully defined after a detailed in vivo study.

2.4. Viscosity

The usually high viscosity of DES or NaDES is the major drawback that can restrict their use as extraction solvents, as it hampers penetration of the solvent in the extraction matrix. Increasing the temperature of the extraction process can lead to a decrease in viscosity, however, this option is not always the ideal choice, as it is energy consuming and some heat-sensitive phytochemicals may not tolerate the elevated temperature.

A simple way to overcome this problem is the addition of a co-solvent in the extraction medium. In most cases, this co-solvent is water, which maintains the green character of the process; however, organic solvents such as methanol have also been used. In this way, the viscosity is lowered and the extraction is facilitated [35,36]. In the recent work of Koutsoukos et al., water was used as a co-solvent for the extraction of phenolic compounds from brown propolis using ChCl/tartaric acid NaDES, with methanol used as the co-solvent for the extraction of carotenoids from apricot pulp and shrimp head by-products using the same NaDES [37].

The concentration of water in the DES–water or NaDES–water mixture affects the efficiency of the extraction, as has been shown by Bi et al. [38], who showed that a mixture of the ChCl/1,4-butanediol NaDES with 35% water is the optimum medium for the extraction of myricetin and amentoflavone from *Chamaecyparis obtusa*. As another indicative example, the work of Zhao et al. [39] indicates that very efficient extraction of rutin from the flower buds of *Sophora japonica* can be successful using the DES ChCl/triethyleneglycol containing 20% water. The researchers studied the viscosities of 20 DES and concluded that the viscosity increases significantly when more hydrogen bonds are possible among the DES constituents.

The amount of added water in a DES is a factor that should be carefully monitored when DES or NaDES are applied as extraction solvents. Dai et al. [40] showed that the viscosity of DES is affected by the water content, and that if more than 50% water is present the hydrogen bond framework of the DES components is destroyed.

Another approach to overcome the problem of high viscosity is to take advantage of the enormous number of possible combinations of natural compounds that can produce NaDES, in order to design solvents of low viscosity. The latest research from Marrucho et al. [41–43] introduced a new concept—the design of less-polar NaDES with lower viscosity, formed by mixing fatty acids of different alkyl chain lengths or by combining menthol with various organic acids.

2.5. Extraction Time

Extraction times show little variance among extractions. Greater extraction times increase costs, while shorter extractions run the risk of leaving considerable quantities of target substances in the sample, rendering the process ineffective. Most of the processes reviewed have very high retrieval percentages, with extraction times ranging from 20 minutes to 2 hours. Naturally, the type of extraction also defines the extraction time necessary, with energy-assisted methods such heating, ultrasound, or microwave requiring less extraction time, but, in turn, more energy to conduct. Overall, the use of DES has enabled undeniably short extraction times for all extraction methods employed.

2.6. Water Content

Water content is another crucial factor to the efficiency of NaDES in flavonoid extractions. Water might be found in a DES system unintentionally (during the synthesis, or from remaining in a container) or intentionally through co-solution to create an aqueous system. While some NaDES could be applied as extraction media on their own, their increased viscosity would hamper the speed of extractions. Furthermore, using pure DES as extraction media could increase the costs of extractions, rendering the process cost-inefficient at the laboratory or industrial scale. To tackle this issue, aqueous solutions of NaDES have been experimentally used instead, attempting the extraction of multiple substances with aqueous NaDES solutions of various concentrations, ranging from 20% all the way up to 80%. According to the literature, particularly in studies in which the water content was a part of optimization, percentages close to 20% water content [20,44] are the balance between creating a fluid extraction system and maintaining the hydrogen bond mesh of the DES [40]. Higher water contents tend to break the hydrogen bond structure of the DES, decreasing its effectiveness. This, however, does not imply that higher water contents would be ineffective or undesirable, as every extraction, having a great number of variables, could take advantage of higher or lower percentages of water.

2.7. DES as Additives, or Additives to DES

DES have been tested both as additives to traditional extraction or analysis systems, or have had additives combined with them in extractions, with ionic liquids being the prevalent example [45]. Sharing similar properties in structure and behavior as solvents meant that the combination of DES and ILs was inevitable.

Additives that co-exist with DES in extractions include, of course, other DES ingredients, which form ternary systems that aim at isolating multiple compounds, enhancing the efficiency of a single extraction or otherwise supplementing the processes which the system will traverse. DES can be made of multiple ingredients, as attempted on multiple occasions, however, there is no guarantee that even a carefully planned and synthesized ternary DES system will be more efficient than a binary, simpler one. Depending on the target, a ternary system could be a better or a worse option [46].

Other additives can be added to assist the extraction of a substance or any other action, however, similarly to the ternary systems, any addition may promote or hamper an extraction, depending on the target. An example by Georgantzi et al. [47] shows that the addition of β-cyclodextrin alters the extractability of select flavonoids depending on the selected DES, benefiting one but worsening the other, with varying levels of significance based on the miscellaneous parameters of the extraction.

2.8. Solvent/Sample Ratio

The ratio of sample (solid or otherwise) to solvent used can also affect the extraction. Immersing a miniscule amount of sample into the solvent means the extraction could be inefficient at a larger scale, since only a small amount of sample is being processed at a given time. On the other hand, smothering the solvent with a copious amount of solid sample might mean the dispersion of solvent around the sample would be slower, the contact surface of the sample with the solvent could eventually decrease (compared to a lesser amount of sample), and the system would end up underperforming. Most of the literature examples, after statistical analysis or reference to previous successful work, have converged on a ratio of solid sample to solvent of 1:10, balancing the amount of sample processed with the efficiency of the method [48,49].

2.9. pH

The pH level of the system may dictate the form of the target compound in some cases, eventually affecting its solubility in the DES. Some DES ingredients might also be affected by the pH themselves, changing their polarity or general behavior, with beneficial or undesirable results. The form of the target (a result of the pH) may completely change the design of the extraction, given that a solvent with a completely different polarity would be extracted by a different DES than originally planned [50,51].

2.10. Separation Techniques

The majority of the research with regards to the use of DES in flavonoid extractions still revolves around the optimization of conditions and evaluating the performance of the extraction at a level deemed satisfactory after having modified certain parameters. Most cited attempts employed HPLC (High Performance Liquid Chromatography), UV-Vis (Ultra-Violet – Visible Spectroscopy) [12], or other instrumental analysis, aiming to analyze the extract in order to evaluate the performance of the

DES mixtures to capture the target compounds. Therefore, some reviews do not capitalize upon the separation of the flavonoids from each other or from a group of organic compounds extracted via the DES as a research point in itself, rather than as a means to examine the performance of the initial extraction. For example, research aiming at simply verifying the flavonoid content of the extract to confirm a successful NaDES extraction employed the Folin–Ciocalteu reagent in a properly prepared extract sample, which, after reaction and incubation, could be studied through UV-Vis or another method to verify the total phenolic content, and subsequently, the flavonoid content [52].

Whenever deemed necessary, often due to the extraction methodology, some researchers opted for filtration or centrifugation prior to the commonplace HPLC analysis that would confirm the extraction of the target flavonoids [53].

A popular and efficient solution for the enrichment and the separation of the flavonoids from the DES extract involves the use of column chromatography through a packed column with a macroporous resin (such as ME-2 [44], NKA-9, or AB-8 [54]), which is cleaned with deionized water, and then after exposure to the extract is eluted with aqueous ethanol. This method, however, requires further processing of the solution to isolate a particular flavonoid from a potential group of flavonoids extracted from a source. The packed column method may provide exceptionally high yields of up to 98.92% [55].

A back extraction using an antisolvent is a simpler method of isolating the flavonoids from the DES. After centrifugation of the sample, the supernatant is diluted with an organic antisolvent such as methanol [46], and subsequently centrifuged again to create a biphasic system. The newly occurring supernatant is the target system, leaving only the solvent to be evaporated (i.e., by vacuum centrifugal evaporation [8]).

Finally, a novel method for extracting the flavonoids from the DES involves the mimicking of DNA denaturation in the DES, as described by Tian et al. [56]. According to their research, the main goal in removing the flavonoids from the DES is the breakdown of the hydrogen mesh that holds the DES together, which is a process similar to denaturation. The group designed an effective method of extracting the flavonoids on a chrome metal organic framework (Materials of Institut Lavoisier: MIL-100 (Cr)), from which the isolation of the flavonoids becomes easier and more selective. Initially, the DES is diluted in water (10% DES Solution), then NaCl is added, causing the HB mesh to breakdown. The subsequent addition of the MIL allows for the readily collectible flavonoids to attach to it and be easily removed from the diluted DES.

An overview of all the applications of NaDES on flavonoid extractions examined in this review is presented in Table 2, including the bioactivity of the target flavonoids as mentioned in each study.

Table 2. Overview of extraction examples from the literature, presenting the plant source, DES employed, target compound, and the bioactive role the writers mention (literature presented in chronological order).

Plant Source	DES	Target Flavonoids	Bioactivity	Ref.
Chamaecyparis obtusa	ChCl (1:2)	Ethyl Glycol Glycerol 1,2-Butanediol 1,3-Butanediol 1,4-Butanediol 2,3-Butanediol 1,6-Butanediol	Myricetin Amentoflavone	Anti-oxidative phenols.	[38]
Radix scutellariae	ChCl (1:2)	1,4-Butanediol Glycerol Ethylene glycol Citric acid Malic acid Lactic acid (also 3:1,2:1,1:1,1:3 1:4) Glucose Sorbitol Sucrose Maltose	Baicalin Wogonoside Baicalein Wogonin	Antiviral, antitumor, anticonvulsant, anti-allergic, anti-inflammatory, anxiolytic, and anti-oxidant properties.	[44]
	Citric Acid (1:2)	Sucrose Glucose			
	Lactic Acid (1:2)	Sucrose			
Chamaecyparis obtusa	ChCl (1:2)	Ethylene Glycol 1,2-Butanediol 1,6-Hexanediol	Quercetin, Myricetin, Amentoflavone	**Myricetin**: Potential anticancer activity and chemoprevention agent for bladder cancer. **Quercetin**: Antibacterial agent that inhibits the oxidation of low-density lipoproteins; anti-allergenic. **Amentoflavone**: Potential cancer growth and metastasis inhibitor; antibacterial, anti-inflammatory, and anti-oxidative.	[48]
	ZnCl$_2$ (1:2)	Ethylene Glycol 1,2-Butanediol 1,6-Hexanediol			
	Et$_4$NCl (1:2)	Phenol Urea Oxalic acid			
	Me(Ph)$_3$PBr (1:4)	1,2-Butanediol Glycerine			
	Me(Ph)$_3$PBr	Ethylene Glycol (1:1,1:2,1:3,1:4,1:5)			

Table 2. *Cont.*

Plant Source	DES		Target Flavonoids	Bioactivity	Ref.
Flos sophorae	ChCl	Glycerol (1:1) Xylitol (5:2) D-(+)-Glucose (1:1)	Rutin Quercetin Kaempferol Isorhamnetin glycosides	**Rutin**: Antiplatelet, anticarcinogen, vasodilatant.	[7]
	L-Proline	D-(+)-Glucose (5:3)			
	Citric Acid	D-(+)-Glucose (1:1) Adonitol (1:1)			
	Betaine	DL-Malic acid (1:1)			
Pyrola incarnata	ChCl	Polyols in different ratios	Hyperin, 2′-O-galloylhyperin Quercitrin Quercetin-O-rhamnoside Chimaphilin	**Hyperin, 2′-O-galloylhyperin**: Anti-inflammatory activity, treatment for cough, blood pressure, and can lower cholesterol; protects the cardiovascular and cerebrovascular networks. **Quercitrin, Quercetin-O-rhamnoside**: Expectorant, cough and asthma ailment, can lower blood pressure and content of fat in the blood, can increase coronary artery blood flow and cause its expansion. **Chimaphilin**: Secondary metabolite of *Pyroloideae crude*, main component of antibacterial, anti-inflammatory. and analgesic products.	[21]
Cajanus Cajan (L) Millsp.	ChCl (1:1)	Sucrose 1,2-Propanediol Glucose Sorbitol Glycol Glycerol 1,3-Butanediol 1,4-Butanediol 1,6-Butanediol (multiple ratios)	Genistin Genistein Apigenin	**Genistein**: Plant estrogen with anti-oxidant, anti-inflammatory, anticarcinogenic, and protein kinase inhibitory action. **Genistin**: A glycoside of genistein with anticancer, antiviral, anti-oxidant, anti-inflammatory, and free radical scavenging potential. **Apigenin**: A naturally occurring flavonoid, with anti-inflammatory action, blood pressure decreasing potential, anti-arteriosclerotic, anti-anxiety, antimicrobial, antiviral, anti-oxidant, and free radical scavenging actions.	[20]
	Glucose	L-Proline Lactic acid			

Table 2. *Cont.*

Plant Source	DES		Target Flavonoids	Bioactivity	Ref.
Equisetum Palustre L.	ChCl	Glycerol 1,4-Butanediol 1,3-Butanediol Ethylene glycol	Nine glycosides	Gastroprotective effect, anti-oxidant, antimicrobial, and genotoxicity activity.	[54]
	Betaine	Glycerol 1,4-Butanediol 1,3-Butanediol Ethylene glycol			
Camelia sinensis leaves	ChCl (1:2)	Ethylene glycol Glycerol 1,4-Butanediol Lactic acid Malic acid Citric acid Glucose Fructose Sucrose	Catechins	Anti-oxidant, antibacterial, antiviral, anti-inflammatory, anti-allergic, anti-hypertensive, anti-obesity, and antidiabetic activity.	[57]
Sophora japonica	ChCl	Urea Acetamide Ethylene glycol Glycerol 1,4-Butanediol Triethylene glycol Xylitol D-Sorbitol p-tolunesulfonic acid Oxalic acid Levulinic acid Malonic acid Malic acid Citric acid Tartaric acid Xylose/water Sucrose/water Fructose/water Glucose/water Maltose/water	Rutin	**Rutin**: Treatment of hypertension and hemostatic for cerebral hemorrhage.	[39]

Table 2. *Cont.*

Plant Source	DES		Target Flavonoids	Bioactivity	Ref.
Virgin Olive Oil	ChCl	Glycerol (1:2) Lactic acid (1:2) Urea (1:2) Sucrose (1:1) Sucrose (4:1) 1,4_Butanediol (1:5) Xylitol (2:1) 1,2_Propanediol (1:1) Malonic acid (1:1) Urea/Glycerol (1:1:1)	Several compounds	**Oleocanthal**: Anti-inflammatory activity, inhibits cyclooxygenase (COX-1 and COX-2) enzymes, similar to the function of ibuprofen (more bioactive roles mentioned). **Oleacein**: Prevents tumor cell proliferation and anti-oxidant properties. **Oleuropein aglycon** (Hy-EA) was found to be an anti-allergenic, to directly regulate HER-2 in breast cancer cells, and fortify against Alzheimer's disease. **Hydroxytyrosol**: Inhibits tumor cell proliferation and promotes apoptosis, broad range of beneficial physiological activities in terms of plasma lipoproteins, oxidative damage, platelet and cellular function, and bone health, due to its anti-inflammatory, antimicrobial, and anti-oxidant activities **Lignans**: Phyto-estrogens; prevention and treatment of cancer, arteriosclerosis, and osteoporosis.	[58]
	D-(+)- Fructose	D-(+)-Glucose/Sucrose (1:1:1)			
Dalbergia odorifera T. Chen leaves	ChCl	Ethylene Glycol Glycerol 1,2-butanediol 1,3-butanediol 1,4-butanediol 2,3-butanediol 1,6-hexylene glycol Lactic acid Citric acid Glucose Sucrose	Prunetin Tectorigenin Genistein Biochanin A	Anticancer, antiviral, anti-oxidant, anti-inflammatory, anti-osteoporosis, cardioprotective, hypoglycemic, anaphylaxis inhibitory activities.	[59]
Ginko biloba	ChCl	Ethylene glycol Glycerol Propylene glycol	Quercetin Myricetin	Anticarcinogenic activity, anti-oxidant, and antiplatelet activities.	[49]

Table 2. *Cont.*

Plant Source	DES		Target Flavonoids	Bioactivity	Ref.
Anredera cordifolia Steenis	Betaine	1,4-butanediol	Vitexin	N/A	[19]
Platycladi Cacumen	ChCl	Laevulinic acid (1:2) Ethylene Glycol (1:2) N,N' Dimethylurea (1:1) -D- Glucose (1:1)	Multiple flavonoids and aglycones	Spasmolytic, antiphlogistic, anti-oxidative, anti-allergenic, and diuretic properties (attributed to the flavonoids of the plant as a whole).	[55]
	Betaine	Laevulinic acid (1:2) Ethylene Glycol (1:2) 1-Methylurea (1:1) -d-Glucose (1:1)			
	L-Proline	Laevulinic acid (1:2) Glycerol (1:2.5) Acetamide (1:1) d-Glucose (1:1)			
Camelia sinensis leaves	Multiple based on betaine, citric acid, and glycerol		Catechins	Anti-oxidant, anticancer, anti-inflammatory, antibacterial, antiviral, and anti-angiogenic properties.	[6]
Ginkgo Biloba leaves	**Hydrophobic phase** Methyl trioctyl ammonium, capryl alcohol, and octylic acid (1:2:3).	**Hydrophilic phase** Ch-Laevulinic acid, Betaine-Etheylene Glycol Ch- Malonic acid	Flavonoids Terpene Trilactones Procyanidine Polyprenyl Acetates	N/A	[60]
Coridothymus capitatus, Origanum vulgare, Salvia fruticosa (triloba), Salvia officinalis, Thymus vulgaris	Lactic Acid (7:1) With/without β-Cyclodextrin	Nicotinamide Ammonium Acetate Sodium Acetate L-Alanine Glycine ChCl	Multiple compounds	Anti-oxidant activity, antimicrobial activity, as well as chemoprotective potency.	[47]
Products containing soy	Seventeen NaDES based on choline chloride and organic acids (e.g., citric acid)		Genistein Daidzein Genistin Daidzin Biochanin A	Isoflavones (the broader category of the target compounds) possess oestrogenic, anti-oxidant, and anti-allergic properties. In the cosmetics industry, they serve to delay the start of skin aging, stimulate collagen biosynthesis in fibroblasts, and accelerate the regeneration of skin cells.	[46]

Table 2. *Cont.*

Plant Source	DES		Target Flavonoids	Bioactivity	Ref.
Herba artemisiae scopariae	ChCl (1:2)	Formic acid Acetic acid Propionic acid Glycerol Urea 1,2-Butanediol 1,4-Butanediol	Rutin Quercetin Scoparone	Anti-inflammatory, antibacterial, and anti-oxidant properties.	[45]
Mentha Piperita L.	Citric Acid	Glycerol (1:2) Xylitol (1:1) D-(+)-Glucose (1:1)	Phenolic compounds	N/A	[53]
	Urea	Glycerol (2:1) Xylitol (2:1) D-(+)-Glucose (2:1)			
Scutellaria Baicalensis Georgi	ChCl (1:1)	Lactic acid Glucose Glycerol 1,4- Butanediol Ethylene Glycol	Baicalin	Blood pressure decrease, detoxifying, antifever action, and reduces the risk of cardiovascular diseases.	[61]
Cyclocarya-paliurus (Batal.) Iljinskaja leaves	ChCl	Glucose (2:1) Citric acid (1:1) Glycerol (1:1) Urea(1:1) Citric Acid/Glycerol (1:1:1) 1,4-Butanediol (1:5) Lactic acid (1:1) Malonic acid (1:1) Malic Acid/Xylosic Alcohol (1:1:1)	Multiple flavonoids	N/A (bioactivity not directly attributed to flavonoids).	[23]
Pollen typhae	ChCl	1,4-Butanediol (1:4) Glucose (1:4) Glycerol (1:4) 1,4-Butanediol/Glycerol (1:2:2) Lactic acid (1:4) Ethylene Glycol (1:4) 1,2-Propanediol (1:4)	Quercetin Kaempferol Isorhamnetin Naringenin	Anti-oxidant, anti-inflammatory, antigenotoxic, antiprotozoal activity.	[22]
	L-proline	Glycerol (4:11)			

Table 2. *Cont.*

Plant Source	DES		Target Flavonoids	Bioactivity	Ref.
Ginkgo Biloba	Multiple DES based on ChCl, betaine, proline, 1,3-Butanediol, 1,2-Propanediol and xylitol with numerous components		Multiple flavonoids	Protection against capillary fragility, anti-inflammatory agents, anti-oxidants, reduction of edema caused by tissue injury, free radical scavengers.	[62]
Orange Peels	ChCl	Glycerol (1:2,1:3,1:4) Ethylene Glycol (1:2,1:3,1:4)	Phenolic anti-oxidants (gallic acid, ferrulic acid, para-coumaric acid)	Antibiotics, antitumoral agents, anti-inflammatory, anti-allergic.	[63]
Moringa Oleifera	ChCl	Citric acid	Flavonoids, among other compounds	Antidepressant, anti-oxidant, antibacterial, antidiabetic, renoprotective, hepatoprotetive, anti-inflammatory, and antilipidemic activities.	[64]
Juglans Regia L.	ChCl (1:2)	Butyric acid Phenylpropionic acid	Quercetin and miscellaneous compounds	N/A	[65]
Sea Buckweed Leeves	ChCl (1:1)	Citric acid Malic acid Lactic acid Ethylene Glycol 1,3-butanediol 1,4-butanediol 1,6-hexanediol 1,2-propanediol Glycerol Glucose Fructose Sucrose	Rutin Quercetin-3-O-glucoside Quercetin Kaempferol Isorhamnetin	Anti-oxidant, cytoprotective, immune-modulatory, cardioprotective, anti-inflammatory, and wound-healing activity.	[10]
Radix Scutellariae	L-Proline	Glycerol(1:4) Glucose/H_2O (5:3:8) Fructose / H_2O (1:1:5)	Scutellarin Baicalin Wogonoside Baicalein Wogonin	Antibacterial and antiviral action.	[66]
	ChCl	Glycol(1:4) Glycerol(1:4) 1,2-Propylene(1:4) 1,2-Butanediol(1:4) Lactic acid(1:4) Malic acid/ H_2O(1:1:3) Glucose/ H_2O(1:1:2)			

Table 2. *Cont.*

Plant Source	DES		Target Flavonoids	Bioactivity	Ref.
Onion solid waste	Sodium Propionate	Lactic acid Glycerol	Quercetin, Rutin	N/A	[52]
Ruta graveolens L.	ChCl (2:1)	Citric acid	Total phenolic content (notably Rutin)	**Rutin**: Inhibition of vascular endothelial growth factor in subtotic concentrations in vitro, angiogenesis inhibitor, support and strengthening of blood vessels, eye strengthener, strong anti-oxidative action.	[67]
Chlorella vulgaris	ChCl	Glycerol (1:2, 1:3, 1:4) Ethylene Glycol (1:2,1:3,1:4) 1,3-Propanediol (1:2,1:3,1:4) 1,4-Butanediol (1:2,1:3,1:4)	Total phenolic content	Anti-oxidant, anti-inflammatory, antimicrobial, and antitumoral properties.	[68]
Five *Gallium* species	NaDES in combination with NaCl (NaDES structure not mentioned) using dispersive liquid–liquid microextraction		Catechin Rutin Quercetin	Total phenolic and flavonoid content, anti-oxidant capacity, and enzyme inhibitory effects of the extracts are mentioned.	[69]
Tomatoes, onions, grapes	Tetrabutylammonium chloride (TBACl) (1:2, 1:3 and 1:4) Tetrabutylammonium bromide (TBABr) (1:3)	Decanoic acid (DA) Decanoic acid (DA)	Quercetin	N/A	[50]
Flos Sophorae	ChCl	Malic acid (1:1, 1:3) Citric acid (1:1, 1:3) Malonic acid (1:1, 1:3) Methylurea (1:1, 1:3) Urea (1:1, 1:3) N,N′-Dimethylurea (1:1, 1:3) 1,2-Butanediol (1:1, 1:3) Ethylene Glycol (1:1, 1:3) Glycerol (1:1, 1:3)	Flavonoids	N/A	[56]

Table 2. *Cont.*

Plant Source	DES		Target Flavonoids	Bioactivity	Ref.
Grapefruit Peels	Lactic Acid	Sodium acetate Ammonium Acetate Glucose Glycine	Polyphenols (notably naringin)	**Polyphenols**: Use in the food industry to prevent the oxidation of lipids. **Naringin**: Food additive towards treating obesity and diabetes (among other uses).	[70]
Flos Sophorae	ChCl	Glycol (1:1) Glycerol (1:1) Sucrose (1:1) 1,3-Butanediol (1:1) 1,4-Butanediol (1:1,2:1,3:1,4:1) Citric acid (1:1) Lactic acid (1:1) Glucose (1:1) Malic acid (1:1)	Rutin nicotiflorin narcissin quercetin kaempferol isorhamnetin	N/A	[71]
Common buckwheat sprouts	ChCl	Acetamide (1:2) Triethylene glycol (1:4) 1,2-Propanediol (1:1) 1,4-Butanediol (1:3) Urea (1:2) Ethylene glycol (1:2) Glycerol (1:1) Oxalic acid (1:1) Malonic acid (1:1)	Orientin Isoorientin Vitexin Isovitexin Quercetin-3-O-robinobioside Rutin	Anti-oxidants.	[72]
Lycium barbarum L. fruits	ChCl	1,2-Propanediol (1:2) Glycerol (1:2) Ethylene Glycol (1:2) Malic acid (1:1) Malonic acid (1:1) p-Toluenesulfonic acid (1:2) Laevulinic acid (1:2) Oxalic acid (1:2) Resorcinol (1:3) Xylitol (1:1) Urea (1:2)	Myricetin Morin Rutin	Anti-oxidant, anticancer, anti-inflammatory, antimicrobial, antiviral, antitumor, neuroprotective properties, enhancement of the lipid metabolism against obesity complications.	[73]
Citrus peel waste	Multiple based on ChCl		Compounds including flavonoids	Anti-oxidant, anti-inflammatory, anticarcinogenic, antiviral, and neuroprotective actions.	[51]
Brown Greek propolis	ChCl	Tartaric acid	Flavonoids and phenolics	Anti-oxidant.	[37]

3. Conclusions

The use of DES in the extractions of flavonoids has yielded overwhelmingly promising results. All of the applications reviewed have reported success in isolation and extraction of the target compounds, as well as competitive, if not superior, extraction rates compared with conventional solvents, in addition to the satisfactory behavior of the extract in the latter applications (such as direct analysis, synthesis or catalysis), wherever attempted.

The issue of selectivity of the DES or NaDES used for the extraction of flavonoids has not been extensively researched yet. In fact, the majority of the published works focus on the extraction efficiency evaluated in terms of total flavonoid or total phenolic content, or anti-oxidant activity, and do not usually analyze the selectivity of the solvents on the extraction of certain molecules. Obviously, this is the next step that should be investigated. As a very good indicative example, Vieira et al. [65] have screened a series of DES comprised of ChCl and carboxylic acids as solvents for the extraction of phytochemicals from the leaves of walnut trees (*Juglans. regia* L.). They found differences in selectivity in the extraction of 3-O-caffeoylquinic acid, quercetin 3-O-glucose, and quercetin O-pentoside among the various DES, and these results enabled them to choose the optimum system for their process.

DES are extremely versatile, with variables such as the number of ingredients, their ratios, and the ingredients themselves. This versatility limits their number to the foresight and the resilience of the researcher, who can design and apply any DES to any system, through any method, with the promise that the resulting extraction will be efficient while being green, given the general properties of DES. The relatively low cost of DES ingredients coupled with an increased selectivity after careful planning means that a lean-process future would definitely involve DES in the extraction compendium.

The fact that the DES extract system has a satisfactory performance in analytical methods [12] or chemical processes means that there is the potential for circumventing the last stage of extraction (separating the extract from the DES), using the system in its entirety instead. DES, being perfectly capable solvents, albeit viscous, can act as the solvent or carrier for the extract in its following stages, in some cases aiding in the reactions it might partake in or in the overall environment the extract will inhabit. The natural origin of NaDES would mean that such systems can be used in their entirety, even in products to be used or consumed by people (such as cosmetics or pharmaceuticals), without the need for extract isolation and further processing.

A pioneer example of NaDES being utilized as both an extraction medium and a biocompatible, consumer-grade carrier is the patented process by Lavaud et.al. [74], which encompasses all of the advantages and the potential they provide. This patent officially links the laboratory research results regarding the use of betaine-based NaDES, mixed with glycerol or water, for the extraction and storage of natural extracts from plants or microorganisms, and the subsequent use of the extract directly as a uniform, natural-origin product. Evidently, the patent of such a method means that the industry is ready and willing to employ NaDES as an immediate extraction-carrier system, even on a consumer level.

This summary of flavonoid extractions via DES incorporates the essence of green chemistry, and the future of chemistry in general. Tailor-made solvents are applied to carefully designed conditions to collect a very valuable substance from a natural source with as little waste as possible. The simplicity of this sentence is deceptive, since careful design and research is required to replace conventional methods, however, these promising results can only act as fuel for future research.

Funding: This research received no external funding.

Conflicts of Interest: The authors declare no conflict of interest.

References

1. Panche, A.N.; Diwan, A.D.; Chandra, S.R. Flavonoids: An overview. *J. Nutr. Sci.* **2016**, *5*, E47. [CrossRef] [PubMed]

2. Babu, P.V.A.; Liu, D. Flavonoids and Cardiovascular Health. In *Complementary and Alternative Therapies and the Aging Population*; Watson, R.R., Ed.; Academic Press: Elsevier: Amsterdam, The Netherlands, 2009; pp. 371–392.
3. Watson, R.; Preedy, V.; Zibadi, S. *Polyphenols in Human Health and Disease*; Academic Press: Elsevier: Amsterdam, The Netherlands, 2014; Volume 1–2.
4. Smith, E.L.; Abbott, A.P.; Ryder, K.S. Deep Eutectic Solvents (DESs) and Their Applications. *Chem. Rev.* **2014**, *114*, 11060–11082. [CrossRef] [PubMed]
5. Weiz, G.; Braun, L.; Lopez, R.; de Maria, P.D.; Breccia, J.D. Enzymatic deglycosylation of flavonoids in deep eutectic solvents aqueous mixtures Paving the way for sustainable flavonoid chemistry. *J. Mol. Catal. B: Enzym.* **2016**, *130*, 70–73. [CrossRef]
6. Jeong, K.M.; Ko, J.; Zhao, J.; Jin, Y.; Yoo, D.E.; Han, S.Y.; Lee, J. Multifunctioning deep eutectic solvents as extraction and storage media for bioactive natural products that are readily applicable to cosmetic products. *J. Clean. Prod.* **2017**, *151*, 87–95. [CrossRef]
7. Nam, M.W.; Zhao, J.; Lee, M.S.; Jeong, J.H.; Lee, J. Enhanced extraction of bioactive natural products using tailor-made deep eutectic solvents:application to flavonoid extraction from Flos sophorae. *Green Chem.* **2015**, *17*, 1718–1727. [CrossRef]
8. Liu, Y.; Garzon, J.; Friesen, J.B.; Zhang, Y.; McAlpine, J.B.; Lankin, D.C.; Chen, S.-N.; Pauli, G.F. Countercurrent assisted quantitative recovery of metabolites from plant-associated natural deep eutectic solvents. *Fitoterapia* **2016**, *112*, 30–37. [CrossRef]
9. Fu, N.; Lv, R.; Guo, Z.; Guo, Y.; You, X.; Tang, B.; Han, D.; Yan, H.; Row, K.H. Environmentally friendly and non-polluting solvent pretreatment of palm samples for polyphenol analysis using choline chloride deep eutectic solvents. *J. Chromatogr. A* **2017**, *1492*, 1–11. [CrossRef]
10. Cui, Q.; Liu, J.-Z.; Wang, L.-T.; Kang, Y.-F.; Meng, Y.; Jiao, J.; Fu, Y.-J. Sustainable deep eutectic solvents preparation and their efficiency in extraction and enrichment of main bioactive flavonoids from sea buckthorn leaves. *J. Clean. Prod.* **2018**, *184*, 826–835. [CrossRef]
11. Gomez, F.J.V.; Espino, M.; de los Angeles Fernandez, M.; Raba, J.; Silva, M.F. Enhanced electrochemical detection of quercetin by Natural Deep Eutectic Solvents. *Anal. Chim. Acta* **2016**, *936*, 91–96. [CrossRef]
12. Paradiso, V.M.; Clemente, A.; Summo, C.; Pasqualone, A.; Caponio, F. Towards green analysis of virgin olive oil phenolic compounds: Extraction by a natural deep eutectic solvent and direct spectrophotometric detection. *Food Chem.* **2016**, *212*, 43–47. [CrossRef]
13. Li, X.; Dai, Y.; Row, K.H. Preparation of two-dimensional magnetic molecularly imprinted polymers based on boron nitride and a deep eutectic solvent for the selective recognition of flavonoids. *Analyst* **2019**, *144*, 1777–1789. [CrossRef] [PubMed]
14. Gao, F.; Liu, L.; Tang, W.; Row, K.H.; Zhu, T. Optimization of the chromatographic behaviors of quercetin using choline chloride-based deep eutectic solvents as HPLC mobile-phase additives. *Sep. Sci. Technol.* **2018**, *53*, 397–403. [CrossRef]
15. Taylor, K.M.; Taylor, Z.E.; Handy, S.T. Rapid synthesis of aurones under mild conditions using a combination of microwaves and deep eutectic solvents. *Tetrahedron Lett.* **2017**, *58*, 240–241. [CrossRef]
16. Fu, N.; Li, L.; Liu, X.; Fu, N.; Zhang, C.; Hu, L.; Li, D.; Tang, B.; Zhu, T. Specific recognition of polyphenols by molecularly imprinted polymers based on a ternary deep eutectic solvent. *J. Chromatogr. A* **2017**, *1530*, 23–34. [CrossRef]
17. Unlu, A.E.; Prasad, B.; Anavekar, K.; Bubenheim, P.; Liese, A. Investigation of a green process for the polymerization of catechin. *Prep. Biochem. Biotechnol.* **2017**, *47*, 918–924. [CrossRef] [PubMed]
18. Roohinejad, S.; Koubaa, M.; Barba, F.J.; Greiner, R.; Orlien, V.; Lebovka, N.I. Negative pressure cavitation extraction: A novel method for extraction of food bioactive compounds from plant materials. *Trends Food Sci. Technol.* **2016**, *52*, 98–108. [CrossRef]
19. Mulia, K.; Muhammad, F.; Krisanti, E. Extraction of vitexin from binahong (*Anredera cordifolia* (Ten.) Steenis) leaves using betaine -1,4 butanediol natural deep eutectic solvent (NADES). *AIP Conf. Proc.* **2017**, *1823*, 020018-1–020018-4.
20. Cui, Q.; Peng, X.; Yao, X.-H.; Wei, Z.-F.; Luo, M.; Wang, W.; Zhao, C.-J.; Fu, Y.-J.; Zu, Y.-G. Deep eutectic solvent-based microwave-assisted extraction of genistin, genistein and apigenin from pigeon pea roots. *Sep. Purif. Technol.* **2015**, *150*, 63–72. [CrossRef]

21. Yao, X.-H.; Zhang, D.-Y.; Duan, M.-H.; Cui, Q.; Xu, W.-J.; Luo, M.; Li, C.-Y.; Zu, Y.-G.; Fu, Y.-J. Preparation and determination of phenolic coumpounds from *Pyrola Incarnata* Fisch. with a green polyols based-deep eutectic solvent. *Sep. Purif. Technol.* **2015**, *149*, 116–123. [CrossRef]

22. Meng, Z.; Zhao, J.; Duan, H.; Guan, H.; Zhao, L. Green and efficient extraction of four bioactive flavonoids from Pollen Typhae by ultrasound-assisted deep eutectic solvents extraction. *J. Pharm. Biomed. Anal.* **2018**, *161*, 246–253. [CrossRef]

23. Shang, X.; Tan, J.-N.; Du, Y.; Liu, X.; Zhang, Z. Environmentally-Friendly Extraction of Flavonoids from *Cyclocarya paliurus* (Batal.) Iljinskaya Leaves with Deep Eutectic Solvents and Evaluation of Their Antioxidant Activities. *Molecules* **2018**, *23*, 2110. [CrossRef] [PubMed]

24. Wang, Y.; Hou, Y.; Wu, W.; Liu, D.; Ji, Y.; Ren, S. Roles of a hydrogen bond donor and a hydrogen bond acceptor in the extraction of toluene from n-heptane using deep eutectic solvents. *Green Chem.* **2016**, *18*, 3089–3097. [CrossRef]

25. Hayyan, M.; Hashim, M.; Hayyan, A.; Al-Saadi, M.A.; Al Nashef, I.M.; Mirghani, M.E.; Saheed, O.K. Are deep eutectic solvents benign or toxic? *Chemosphere* **2013**, *90*, 2193–2195. [CrossRef] [PubMed]

26. Paiva, A.; Craveiro, R.; Aroso, I.; Martins, M.; Reis, R.L.; Duarte, A.R.C. Natural deep eutectic solvents-solvents for the 21st century. *ACS Sustain. Chem. Eng.* **2014**, *2*, 1063–1071. [CrossRef]

27. Ahmadi, R.; Hemmateenejad, B.; Safavi, A.; Shojaeifard, Z.; Mohabbati, M.; Firuzi, O. Assessment of cytotoxicity of choline chloride-based natural deep eutectic solvents against human HEK-293 cells: A QSAR analysis. *Chemosphere* **2018**, *209*, 831–838. [CrossRef] [PubMed]

28. Radošević, K.; Bubalo, M.C.; Srček, V.G.; Grgas, D.; Dragičević, T.L.; Redovniković, I.R. Evaluation of toxicity and biodegradability of choline chloride based deep eutectic solvents. *Ecotoxicol. Environ. Saf.* **2015**, *112*, 46–53. [CrossRef]

29. Radošević, K.; Čanak, I.; Panić, M.; Markov, K.; Cvjetko Bubalo, M.; Frece, J.; Gaurina Srček, V.; Radojčić Redovniković, I. Antimicrobial, cytotoxic and antioxidative evaluation of natural deep eutectic solvents. *Environ. Sci. Pollut. Res.* **2018**, *25*, 14188–14196. [CrossRef] [PubMed]

30. Macário, I.P.E.; Oliveira, H.; Menezes, A.C.; Ventura, S.P.M.; Pereira, J.L.; Gonçalves, A.M.M.; Coutinho, J.A.P.; Gonçalves, F.J.M. Cytotoxicity profiling of deep eutectic solvents to human skin cells. *Sci. Rep.* **2019**, *9*, 3932. [CrossRef]

31. Hayyan, M.; Looi, C.Y.; Hayyan, A.; Wong, W.F.; Hashim, M.A. In vitro and in vivo toxicity profiling of ammonium-based deep eutectic solvents. *PLoS ONE* **2015**, *10*, 2. [CrossRef]

32. Mbous, Y.P.; Hayyan, M.; Wong, W.F.; Looi, C.Y.; Hashim, M.A. Unraveling the cytotoxicity and metabolic pathways of binary natural deep eutectic solvent systems. *Sci.Rep.* **2017**, *7*, 41257. [CrossRef]

33. Chen, J.; Wang, Q.; Liu, M.; Zhang, L. The effect of deep eutectic solvent on the pharmacokinetics of salvianolic acid B in rats and its acute toxicity test. *J. Chromatogr. B* **2017**, *1063*, 60–66. [CrossRef] [PubMed]

34. Belebna, M.; Ruesgas-Ramón, M.; Bonafos, B.; Fouret, G.; Casas, F.; Coudray, C.; Durand, E.; Cruz Figueroa-Espinoza, M.; Feillet-Coudray, C. Toxicity of natural deep eutectic solvent betaine: Glycerol in rats. *J. Agric. Food Chem.* **2018**, *66*, 6205–6212. [CrossRef] [PubMed]

35. Alcalde, R.; Atilhan, M.; Aparicio, S. On the properties of (choline chloride+ lactic acid) deep eutectic solvent with methanol mixtures. *J. Mol. Liq.* **2018**, *272*, 815–820. [CrossRef]

36. Ruesgas-Ramón, M.; Figueroa-Espinoza, M.C.; Durand, E. Application of deep eutectic solvents (DES) for phenolic compounds extraction: Overview, challenges, and opportunities. *J. Agric. Food Chem.* **2017**, *65*, 3591–3601. [CrossRef] [PubMed]

37. Koutsoukos, S.; Tsiaka, T.; Tzani, A.; Zoumpoulakis, P.; Detsi, A. Choline Chloride and Tartaric Acid, a Natural Deep Eutectic Solvent for the Efficient Extraction of Phenolic and Carotenoid Compounds. *J. Clean. Prod.* **2019**, in press. [CrossRef]

38. Bi, W.; Tian, M.; Row, K.H. Evaluation of alcohol-based deep eutectic solvent in extraction and determination of flavonoids with response surface methodology optimization. *J. Chromatogr. A* **2013**, *1285*, 22–30. [CrossRef] [PubMed]

39. Zhao, B.-Y.; Xu, P.; Yang, F.-X.; Wu, H.; Zong, M.-H.; Kou, W.-Y. Biocompatible Deep Eutectic Solvents Based on Choline Chloride: Characterization and Application to the Extraction of Rutin from *Sophora japonica*. *ACS Sustain. Chem. Eng.* **2015**, *3*, 2746–2755. [CrossRef]

40. Dai, Y.; Witkamp, G.-J.; Verpoorte, R.; Choi, Y.H. Tailoring properties of natural deep eutectic solvents with water to facilitate their applications. *Food Chem.* **2015**, *187*, 14–19. [CrossRef]

41. Ribeiro, B.D.; Florindo, C.; Iff, L.C.; Coelho, M.A.; Marrucho, I.M. Menthol-based eutectic mixtures: Hydrophobic low viscosity solvents. *ACS Sustain. Chem. Eng.* **2015**, *3*, 2469–2477. [CrossRef]
42. Florindo, C.; Branco, L.; Marrucho, I. Development of hydrophobic deep eutectic solvents for extraction of pesticides from aqueous environments. *Fluid Phase Equilib.* **2017**, *448*, 135–142. [CrossRef]
43. Florindo, C.; Romero, L.; Rintoul, I.; Branco, L.C.; Marrucho, I.M. From phase change materials to green solvents: Hydrophobic low viscous fatty acid–based deep eutectic solvents. *ACS Sustain. Chem. Eng.* **2018**, *6*, 3888–3895. [CrossRef]
44. Wei, Z.-F.; Wang, X.-Q.; Peng, X.; Wang, W.; Zhao, C.-J.; Gang Zu, Y.; Fu, Y.-J. Fast and Green extraction and separation of main bioactive flavonoids from Radix Scutellariae. *Ind. Crop. Prod.* **2015**, *63*, 175–181. [CrossRef]
45. Ma, W.; Row, K.H. Optimized extraction of bioactive compounds from Herba Artemisiae Scopariae with ionic liquids and deep eutectic solvents. *J. Liq. Chromatogr. Relat. Technol.* **2017**, *40*, 459–466. [CrossRef]
46. Bajkacz, S.; Adamek, J. Evaluation of new natural deep eutectic solvents for the extraction of isoflavones from soy products. *Talanta* **2017**, *168*, 329–335. [CrossRef]
47. Georgantzi, C.; Lioliou, A.-E.; Paterakis, N.; Makris, D.P. Combination of Lactic Acid-based Deep Eutectic Solvents (DES) with β-cyclodextrin: Performance screening using ultrasound-assisted extraction of Polyphenols from selected native Greek medicinal plants. *Agronomy* **2017**, *7*, 54. [CrossRef]
48. Tang, B.; Park, H.E.; Row, K.H. Simultaneous Extraction of Flavonoids from *Chamaecyparis obtusa* Using Deep Eutectic solvents as Additives of Conventional Extractions Solvents. *J. Chromatogr. Sci.* **2015**, *53*, 836–840. [CrossRef]
49. Tang, W.; Li, G.; Chen, B.; Zhu, T.; Row, K.H. Evaluating ternary deep eutectic solvents as novel media for extraction of flavonoids from *Ginkgo biloba*. *Sep. Sci. Technol.* **2017**, *52*, 91–99. [CrossRef]
50. Kanberoglu, G.S.; Yilmaz, E.; Soylak, M. Application of Deep Eutectic Solvent in ultrasound-assisted emulsification microextraction of quercetin from some fruits and vegetables. *J. Mol. Liq.* **2019**, *279*, 571–577. [CrossRef]
51. Xu, M.; Ran, L.; Chen, N.; Fan, X.; Ren, D.; Yi, L. Polarity-dependent extraction of flavonoids from citrus peel waste using a tailor-made deep eutectic solvent. *Food Chem.* **2019**, *297*, 124970. [CrossRef]
52. Stefou, I.; Grigorakis, S.; Loupassaki, S.; Makris, D.P. Development of sodium propionate-based deep eutectic solvents for polyphenol extraction from onion solid wastes. *Clean Technol. Environ. Policy* **2019**. [CrossRef]
53. Jeong, K.M.; Jin, Y.; Yoo, D.E.; Han, S.Y.; Kim, E.M.; Lee, J. One-step sample preparation for convenient examination of volatile monoterpenes and phenolic compounds in peppermint leaves using deep eutectic solvents. *Food Chem.* **2018**, *251*, 69–76. [CrossRef] [PubMed]
54. Qi, X.-L.; Peng, X.; Huang, Y.-Y.; Li, L.; Wei, Z.-F.; Zu, Y.-G.; Fu, Y.-J. Green and efficient extraction of bioactive flavonoids from *Equisetum palustre* L. by deep eutectic solvents-based negative pressure cavitation method combined with macroporous resin enrichment. *Ind. Crop. Prod.* **2015**, *70*, 142–148. [CrossRef]
55. Zhuang, B.; Dou, L.-L.; Li, P.; Liu, E.-H. Deep eutectic solvents as green media for extraction of flavonoid glycosides and aglycones from Platycladi Cacumen. *J. Pharm. Biomed. Anal.* **2017**, *134*, 214–219. [CrossRef] [PubMed]
56. Tian, H.; Wang, J.; Li, Y.; Bi, W.; Chen, D.D.Y. Recovery of Natural Products from Deep Eutectic Solvents by Mimicking Denaturation. *ACS Sustain. Chem. Eng.* **2019**, *7*, 9976–9983. [CrossRef]
57. Li, J.; Han, Z.; Zou, Y.; Yu, B. Efficient extraction of major catechins in *Camellia sinensis* leaves using green choline chloride-based deep eutectic solvents. *RSC Adv.* **2015**, *5*, 93937–93944. [CrossRef]
58. Garcia, A.; Rodriguez-Juan, E.; Rodriguez-Gutierrez, G.; Rios, J.J.; Fernandez-Bolanos, J. Extraction of phenolic compounds from virgin olive oil by deep eutectic solvents (DESs). *Food Chem.* **2016**, *197*, 554–561. [CrossRef] [PubMed]
59. Li, L.; Liu, J.-Z.; Luo, M.; Wag, W.; Huang, Y.-Y.; Efferth, T.; Wang, H.-M.; Fu, Y.-J. Efficient extraction and preparative separation of four main isoflavonoids from *Dalbergia odorifera* T. Chen leaves by deep eutectic solvents-based negative pressure cavitation extraction followed by macroporous resin column chromatography. *J. Chromatogr. B* **2016**, *1033–1034*, 40–48. [CrossRef] [PubMed]
60. Cao, J.; Chen, L.; Li, M.; Cao, F.; Zhao, L.; Su, E. Two-phase systems developed with hydrophilic and hydrophobic deep eutectic solvents for simultaneously extracting various bioactive compounds with different polarities. *Green Chem.* **2018**, *20*, 1879–1886. [CrossRef]

61. Wang, H.; Ma, X.; Cheng, Q.; Wang, L.; Zhang, L. Deep Eutectic Solvent—Based ultra-high pressure extraction of Baicalin from *Scutellaria baicalensis* Georgi. *Molecules* **2018**, *23*, 3233. [CrossRef]

62. Yang, M.; Cao, J.; Cao, F.; Lu, C.; Su, E. Efficient Extraction of Bioactive Flavonoids from Ginkgo biloba Leaves Using Deep Eutectic Solvent/Water Mixture as Green Media. *Chem. Biochem. Q.* **2018**, *32*, 315–324. [CrossRef]

63. Ozturk, B.; Parkinsons, C.; Gonzalez-Miquel, M. Extraction of polyphenolic antioxidants from orange peel waste using deep eutectic solvents. *Sep. Purif. Technol.* **2018**, *206*, 1–13. [CrossRef]

64. Hamany Djande, C.Y.; Piater, L.A.; Steenkamp, P.A.; Madala, N.E.; Dubery, I.A. Differential extraction of phytochemicals from the multipurpose tree, *Moringa oleifera*, using green extraction solvents. *South. Afr. J. Bot.* **2018**, *115*, 81–89. [CrossRef]

65. Vieira, V.; Prieto, M.A.; Barros, L.; Coutinho, J.A.P.; Ferreira, I.C.F.R.; Ferreira, O. Enhanced extraction of phenolic compounds using choline chloride based deep eutectic solvents from *Juglans regia* L. *Ind. Crop. Prod.* **2018**, *115*, 261–271. [CrossRef]

66. Xiong, Z.; Wang, M.; Guo, H.; Xu, J.; Ye, J.; Zhao, J.; Zhao, L. Ultrasound-assisted deep eutectic solvent as green and efficient media for the extraction of flavonoids from *Radix scutellariae*. *NJC* **2019**, *43*, 644–650. [CrossRef]

67. Pavic, V.; Flacer, D.; Jakovljevic, M.; Maja, M.; Jokic, S. Assessment of total phenolic content in vitro antioxidant and antibacterial activity of *Ruta graveolens* L. Extracts obtained by choline chloride based natural deep eutectic solvents. *Plants* **2019**, *8*, 69. [CrossRef] [PubMed]

68. Mahmood, W.M.A.; Lorwirachsutee, A.; Theodoropoulos, C.; Gonzalez-Miquel, M. Polyol-Based Deep Eutectic Solvents for Extraction of Natural Polyphenolic Antioxidants from *Chlorella vulgaris*. *ACS Sustain. Chem. Eng.* **2019**, *7*, 5018–5026. [CrossRef]

69. Mocan, A.; Diuzheva, A.; Badarau, S.; Cadmiel, M.; Andruch, V.; Carradori, S.; Campestre, C.; Tartaglia, A.; De Simone, M.; Vodnar, D.; et al. Liquid Phase and Microwave-Assisted Extractions for Multicomponent Phenolic Pattern Determination of Five Romanian Galium Species Coupled with Bioassays. *Molecules* **2019**, *24*, 1226. [CrossRef]

70. El Kantar, S.; Rajha, H.N.; Boussetta, N.; Voroblev, E.; Maroun, R.G.; Louka, N. Green extraction of polyphenols from grapefruit peels using high voltage electrical discharges deep eutectic solvents and aqueous glycerol. *Food Chem.* **2019**, *295*, 165–171. [CrossRef]

71. Wang, G.; Cui, Q.; Yin, L.-J.; Zheng, X.; Gao, M.-Z.; Meng, Y.; Wang, W. Efficient extraction of flavonoids from Flos Sophorae Immaturus by tailored and sustainable deep eutectic solvent as green extraction media. *J. Pharm. Biomed. Anal.* **2019**, *170*, 285–294. [CrossRef]

72. Mansur, A.R.; Song, N.-E.; Jang, H.W.; Lim, T.-G.; Yoo, M.; Nam, T.G. Optimizing the ultrasound-assisted deep eutectic solvent extraction of flavonoids in common buckwheat sprouts. *Food Chem.* **2019**, *293*, 438–445. [CrossRef]

73. Ali, M.C.; Chen, J.; Zhang, H.; Li, Z.; Zhao, L.; Qiu, H. Effective extraction of flavonoids from *Lycium barbarum* L. fruits by deep eutectic solvents-based ultrasound-assisted extraction. *Talanta* **2019**, *203*, 16–22. [CrossRef] [PubMed]

74. Lavaud, A.; Laguerre, M.; Birtic, S.; Fabiano Tixier, A.S.; Roller, M.; Chemat, F.; Bily, A.C. Eutectic Extraction Solvents, Extraction Methods by eutecticgenesis using said solvents, and extracts derived from said extraction methods. U.S. Patent WO 2016/162703, 13 October 2016.

MDPI

St. Alban-Anlage 66

4052 Basel

Switzerland

Tel. +41 61 683 77 34

Fax +41 61 302 89 18

www.mdpi.com

Applied Sciences Editorial Office

E-mail: applsci@mdpi.com

www.mdpi.com/journal/applsci